Insect Predators and Pest Management

ABOUT THE AUTHORS

Miss Vaishali Jayashingrao Patil (M.Sc., Ph.D.) working as contributory teacher to M.Sc. in Department of Agrochemicals & Pest Management, Shivaji University, Kolhapur since 2000. She has published 13 research papers in various scientific journals and actively participated in several national and international conferences/seminars, etc. She visited Indonesia in April, 2002 for plenery lecture on "Wild Silk Moth Diversity in Maharashtra, India" in 4th International Conference on Wild Silk Moths.

Dr. Sathe Tukaram Vithalrao (M.Sc., Ph.D., Sangit Visharad, IBT (Seri.), F.S.E.Sc.) is reader in Entomology, Department of Zoology, Shivaji University, Kolhapur since 1996. He is contributory teacher to M.Sc. courses in Sericulture and Agrochemicals and Pest Management, Shivaji University, Kolhapur since last 7 and 11 years respectively. He has published 130 research papers in the journals of national and international repute, and several articles in daily newspaper/All India radio. He has published 9 books, Preet Patangachi (Drama) (1981), Crop Protection from Insects (1991), Sericultural Crop Protection (1998), Biological Pest Control (2000), Insect Pest Predators (2001), Cotton Pests and Biocontrol Agents (2001), Sericulture and Pest Management (2001), Mosquitoes and Diseases (2002), Indian Pest Parasitoids (2003) and Agrochemicals and Pest Management (2003). He has completed 3 major research projects, produced 11 Ph.D. and 2 M.Phil. students. He visited Canada British Columbia University Vancouver in 1988, Thailand QSNCC Bangkok, 2002 and Japan in 1988 for academic activities. He is member of editorial board of several journals in India. He also contributed for several multi-author books and involved in television programming of butterflies.

Insect Predators and Pest Management

Vaishali J. Patil
Department of Agrochemicals & Pest Management
Shivaji University
KOLHAPUR

&

T.V. Sathe
Entomology Division
Department of Zoology, Shivaji University
KOLHAPUR

2019

Daya Publishing House©
A Division of
Astral International Pvt. Ltd.
New Delhi - 1100 002

First Published, 2003

Reprinted, 2019

ISBN 978-81-7035-319-5 (Hardbound)

Published by : **Daya Publishing House®**
A Division of
Astral International Pvt. Ltd.
– ISO 9001:2015 Certified Company –
4736/23, Ansari Road, Darya Ganj
New Delhi-110 002
Ph. 011-43549197, 23278134
E-mail: info@astralint.com
Website: www.astralint.com

Laser Typesetting : **Classic Computer Services**
Delhi - 110 035

Printed at : **Replika Press Pvt. Ltd.**

PRINTED IN INDIA

Preface

In pest management, chemicals are extensively used. The sole reliance on pesticides create a lot of problems such as pesticide resistance in pests, pest resurgence, secondary pest out break, increasing cost of pesticides, pesticide residues in food, ill effect on non-target organisms and environmental pollution, etc. These maladies of injudicious use of pesticides have forced workers to turn towards biological control as an environmental friendly method of pest control. Biological control now forms an integral part of pest management and alternative for chemical control.

Predators are widely used in biological pest control programmes as they kill various kinds of insect pests. Hence, in the present book attempts have been made on various aspects of predators such as taxonomical description, seasonal abundance, habitats, biology, rearing techniques and predator-prey index, etc. We feel that the book will find interesting, stimulatory guide to students, teachers, farmers and researchers in the field of biological pest control.

Vaishali J. Patil

T.V. Sathe

Contents

Chapter 1
Introduction

A very large number of pests inflict the damage to the stems, leaves, flowers, fruits, seeds and roots of agricultural plants. Many soil inhibiting insects damage roots and kill the plants causing serious losses in the agricultural yield. As per the estimate made in 1970, annual crop losses exceeding Rs. 6,000 Crores are inflected by pests, of which insects cause 20% (Paroda, 1993).

The protection of crops from various kinds of pest remains a unsolved problem and is intensified on certain high yielding hybrid varieties. Now-a-days, several techniques and tactics are adopted in IPM (Integrated Pest Management). The management strategy for each type of pest being different, control operations are complex and highly technical. IPM involves to take a series of steps for keeping the pest population below the level of economic threshold which demand a judicious use of pesticides, disease free seeds of resistant varieties, biological suppression of pest population, usage of pheromones, timely spray of pesticides, etc. According to Smith (1980) IPM is decision making process to determine it, when, where and what strategy and/or mix of tactics should be used.

Pesticide usage is comparatively not new in India. From a merely 4000 metric tonnes of technical grade material in 1950, the usage has grown at a rate of 7.5% per annum and exceed 82 metric tonnes in the year 1991 (Paroda, 1993). It is observed that, insecticides when spread on crops are reported to reduce the populations of phytophagous insects. However, several insecticides, used on orchard and field crops, bring about tremendous increase of the insects against which these have been applied or of other phytophagous species present only in small number at the time of treatment. From the view of over kill of pesticides, the approach has moved towards ecologically sound integrated pest management and biological control is one of the tool of pest control strategy. Pesticides increase the yield of crops but the indiscriminate use of them invite the pest problems in a severe form and total devastation of crops have been noticed. Improper use of pesticides leads to serious problems like air and water pollution, health hazards, physical and physiological changes in the soil, detrimental effects on beneficial insects (predators, parasitoids, honeybees, etc.), destruction of natural balance and ecological cycles, development of resistant in pests, pest resurgence, secondary pest outbreak, etc. The above facts indicates that there is an urgent need to develop an alternative or newer approach to manage the pests that flare-up when the ecosystems are distributed due to the aggressive and unscientific methods (pesticides) adopted by farmers. Thus, it is realized that, biological control is living weapon over chemical control and is modern and prestigious method of pest management adopted at global level.

DeBach (1964) defined biological control as, "the action of parasitoids, predators and pathogens in maintaining other organisms density at a lower average than would occur in their absence". While, Coppel and Mertins (1977) defined as, "the use or encouragement by man of living organisms or their productions for the population reduction of pest insects of biological control".

Biological control is designed with predators and parasitoids. The predators are lions, tigers and hawks of the insect world and act as potent bio-control agents. The predators are defined as, "an organism which directly attack, kills and eats one of the other species (preys)".

Generally predators are larger than their prey, they consume more than one prey individual by moving around freely in both, their immature and adult stages. Predators are categorized in two groups by taking into account of their feeding habit I) Predators stinging and biting II) Predators sucking.

The predators are scattered in about 167 families of 14 orders of Class Insecta. Coleoptera, Neuroptera, Hymenoptera, Diptera and Hemiptera contain a very large number of predatory insects which frequently suppress the pest populations of agricultural crops. Among the predaceous orders, Coleoptera ranks first. The family

Coccinellidae of the order Coleoptera is exclusively predaceous except sub family Epilachninae (Epilachna beetles). Coccinellids / lady bugs / lady birds or preferably lady beetles have been respected through the centuries as predators from this family which contains about 4000 species in the world (Sathe and Bhosale, 2001).

World organizations like CIBC, CAB, CABI, IOBC etc. are actively engaged in doing research in biological pest control. These organizations provide effective services to world agriculture to coordinate and administer agricultural informations, identification, biological control and other services. Manjunath (1992) says that, the commercial insectaries have made remarkable progress in these aspects and are largely responsible for transferring biological technology from lab to land in several countries. In U.S.A. commercial production of predators and parasitoids seems to be quite popular. There are at least 50 commercial insectaries, which annually produce some 203 million *Aphytis melinus* DeBach, 26 million *Cryptolaemus montrouzieri* Mulsant and 18 million *Chrysoperla carnea* Stephens for control of various lepidopteran pests, scale insects, mealy bugs, aphids and other soft bodied sucking pests. In Russia, more than 10 biological factories are engaged in mass production of *Trichogramma* spp., and about 15 species of parasitoids and predators are produced and released in about 10 million hectares for control of several agricultural and horticultural pests. In China, several species of predators are mass reared and used against the pests of various crops, of which cotton alone comprised some 6,80,000 hectare annually. In Briton 170 million *Encarsia* are produced annually for control of white flies infesting vegetable crops in glass houses.

DeBach (1964) reported more than 225 cases of biological control programmes of about 110 species of pests in 60 countries. Biological pest control of insects in South East Asia and the pacific region have been reviewed by Rao *et al.* (1971). A predator, Vedalia Lady bird beetle, *Rodolia cardinalis* (Mulsant) was the first example of insect natural enemy of deliberate manipulation into the California in 1988 to control the cottony cushion scale, *Icerya purchasi* (Maskell) on citrus. According to Sweetman (1958) one third of the successful biological insect pest control programmes were attributed by introduction and release of insect predators.

In 1957, CIBC (Common Wealth Institute of Biological Control) was established at Bangalore, India for giving fillip to Indian

Scientists to work on biological control. Since then a number of centres have been launched in our country. Notable among them are the central biological control station at Gorakhpur (U.P.), Solan (H.P.), Hyderabad (A.P.), Sriganganagar (Rajasthan), Indian Agricultural Research Institute, Pusa (New Delhi), International Crop Research Institute for the Semi Arid Tropics, Patanchery (A.P.), Biocontrol Research Laboratory, Chengalpattu (A.P.), Annamalai University, Annamalainagar (T.N.), Marathwada University, Aurangabad, Shivaji University, Kolhapur, G.K.V.K., Bangalore, E.R.I., Madras, VSI, Pune etc.

CIBC Bangalore have made extensive survey of important natural enemies of aphids, rhinoceros beetles, coconut caterpillars and several other insect pests in India and some of the biocontrol agents were utilized in biological pest control programmes.

C. montrouzieri was the first introduction of predator biocontrol agent in India in 1898 from Australia to control *Coccus viridis* (Green) (Puttarudriah *et al.,* 1952; Mayne, 1953). Later, in 1927 lady bird beetles were introduced to control aphids on lentil at Pusa (Dutt, 1927). The pest occupied various parts of the country including Mysore, Coorg, Travancore and Bombay. The citrus industry in several states was seriously threatened. A substation for multiplication and liberation of *Rodolia* was set up at Kodaikanal with stocks obtained from Nilgiris and New Zealand. Thousands of adults were released for several years in small batches and at intervals, resulting in effective control of the sucking type of pests (Subramanyam, 1955). *C. montrouzieri* have been reared on the Cochineal insect *Dactylopius opuntiae* (Ckll.) in the laboratory and successfully used to control *Pulvinaria maxima* (Green) on neem, *Centrococcus insolitus* (Green) on brinjal, *P. psidii* (Mask) on guava and sapota and *Pseudococcus corymbatus* (Green) on citrus in South India (Rao and David, 1958) and recently against mulberry mealy bug, *Maconellicoccus hirsutus* (Sathe, 1998; Sathe & Bhoje, 2000).

From India *R. cardinalis* and *Rodolia* sp. were sent to Portugal in 1961 and in North West Africa in 1963 for the control of *I. purchasi*. The attempt gave successful results (Simmonds, 1970). The cocinellids *C. montrouzieri, Nephus* sp. and *Hyperaspis* sp. introduced from India to the Bahamas and Barbados in 1968-69 for the control of soft scales on guava and *Saccharicoccus* sp. (pink mealy bug) on sugarcane gave good results. The review of literature (Rao *et al.,*

1971, Sathe, 1998, Sathe & Bhosale, 2001, Sathe & Margaj, 2001, Sathe & Jadhav, 2001 etc.) indicates that, several predators have been introduced for biological pests suppression in various parts of India.

The family Coccinellidae is extremely important from the view of biological control of insect pests. Thousands of coccinellid predators are interacting with pests by keeping the population below the level of economic damage, hence they have been utilized extensively in biological control programme. The Coccinellid predators are found feeding on Homopterans such as aphids, Psyllids, Coccids, mealy bugs and non-insects like mites.

Most of the work on coccinellids in past is devoted to their seasonal abundance, behaviour, preys and utilization in biocontrol etc. but very little attention is paid on their biology and taxonomy.

In past, taxonomical studies of Coccinellids have been attempted by following workers Linnaeus (1758), Fabricius (1775), Mulsant (1866), Lewis (1873, 1879), Crotch (1873), Casey (1899, 1908; Korschefsky, 1931, 1932, 1933), Timberlake (1943), Kapur (1947, 1948a, 1948b, 1950, 1954, 1967, 1969, 1970, 1972), Rafiq Ahmed (1973), Chapian (1974), Ghorpade (1974), Miyatake (1985), Bhaskar (1992), Omkar & Bind (1993) Bhosale (1994) and Sathe & Bhosale, (2001) etc.

Tiger beetles and ground beetles are also predaceous insects which belongs to order Coleoptera of the families cicinelidae and carabidae respectively. Cicindelids are generally recognizable in the field from their general form, which is distinguish from that of their allies, the carabids. They are brightly coloured green, brown or black with spots or bands of white being most common. The majority are from one half to an inch long, few under or over these limits. The head is short and thickset, constricted behind the eye into a neck, eyes are prominent, antennae moderately long. The life history of these beetles is believed to be uniform. The larvae found in vertical burrows in the wet sand or mud near rivers. They require wet material which admits of the formation of a burrow, but their choice of locality may be determined by their prey. The burrow extends vertically from the surface and the larva can more up and down by means of the legs and a dorsal hump or projection. The head is flat, used to carry up the soil when excavating and the very long jaws are turned backwards and upwards, so that when the flat head is blocking the upper end of the tunnel, the jaws have free play above and are in a position to seize any unway insect that alights or walks within reach.

The length of the life-history has not been ascertained but as each species appears to emerge in the imago form for a definite period in the year, it is probably that the life history occupies, one year or multiples of one year, the imago lives for several weeks. The curious larva of *Collyris emarginatus* Deg. from Java, which lives in burrows in coffee stems, feeding on the insects that go past. The larva has on the fifth abdominal segment six hooks, curved forwards, on a protruberance. A similar larva was found in China.

The majority of beetles appear in the rainy months. Some at the beginning, some later. Most of species are diurnal in habit, though some are known to be nocturnal. They are among the most active of insects, flying for short distances with great rapidity and also running quickly. So far as known all are predaceous on other insects. Maindron records that *Derocrania longesulcata*, Mon., feeds on *Silis* (Drilinae), and such records of food are noticeable for their rarity. The majority are found in damp places in rice fields or thick vegetation, on river banks, on the seashore; some are found only on trees in forest localities.

Carabid beetles are distinguished from all others, the only family with which they are confused is the cicinelidae which have the lateral extension of the clypeus in front of the antennae. The family carabidae and cicindelidae have close morphological relationships. The carabid beetles vary in size from small to moderately large, the smallest one is quarter of an inch long, the largest is nearly one inch in length. The colouring is varied from brown to black sometimes with bright batches of yellow and is often strikingly warning. Carabid beetles are characterized by having oval body size broader than cicinelids and are more flattened. The antennae are filiform are rarely moniliform, compound eyes are large.

The life history of predaceous ground beetles is almost unknown in detail but so far as is known elsewhere it is uniform through out the group and little is known about the Indian species. The larvae are slender active insects with large head and long mandibles and six ocelli. The thorax and abdomen are smooth and tapering, with terminal pair of dorsal cerci and anal tube and 3 pairs of thoracic legs. The larvae are the main predaceous individuals which constitute a part of the surface fauna and best found when caterpillars are abundant on the crop. The larvae are exclusive predaceous one in Indian fauna. They suck out the blood of caterpillars and other insects. No details are available as to the length of the life history, however it is known that pupation takes place in soil. The eggs may be large oval bodied, white and soft and vary with the species with respect to size. Hibernation takes place in imago stage, the beetles burying themselves in the soil or otherwise taking shelter .

Carabidae are partly diurnal and nocturnal. In some beetles they have defense arrangement in the form of certain glands by that they can repelled the enemies and protect themselves. No species is in India known to be destructive to agricultural crops. The carnivorous habits of the family apparently giving place to herbivorous habits in a very few species. The family can be classified among the great number of miscellaneous predators which check the general increase of other insects. They are protected by their ferocious habits, their hardness and by a volatile and offensive fluid (Pheropsophus). Their habitat includes every part of the earths surface and they are among the most universal of insects. Many times in the plains, some in the hills, cultivated areas harbour many as do the wastelands and jungles.

Praying mantids is also an important group of predaceous insects. The praying mantids belongs to order Mantoidea which contain about 2000 species in the world. Praying mantids are medium sized, with elongate prothorax, large spined grasping front legs (raptorial). The middle and hind legs are usually slender. The mantids may be long winged, short winged or wingless. They may be green, brown or mottled or in few cases brighter coloured and some have definite patterns. Praying mantids are exclusively, all predaceous species. They feed on other various insects such as flies grasshoppers, caterpillars, butterflies, leafhoppers mealy bugs and some other insects. They also feed on small animals like frogs, lizards etc. Praying mantids are potential pest biocontrol agents however very little attention is paid on this important group, useful in pest management.

In the present text, taxonomical studies have been made on 8 species of coccinellid beetles (lady bird beetles), 4 species of carabid beetles (Tiger beetles) and 9 species of praying mantids. In addition, survey and surveillance studies have been made on three species of

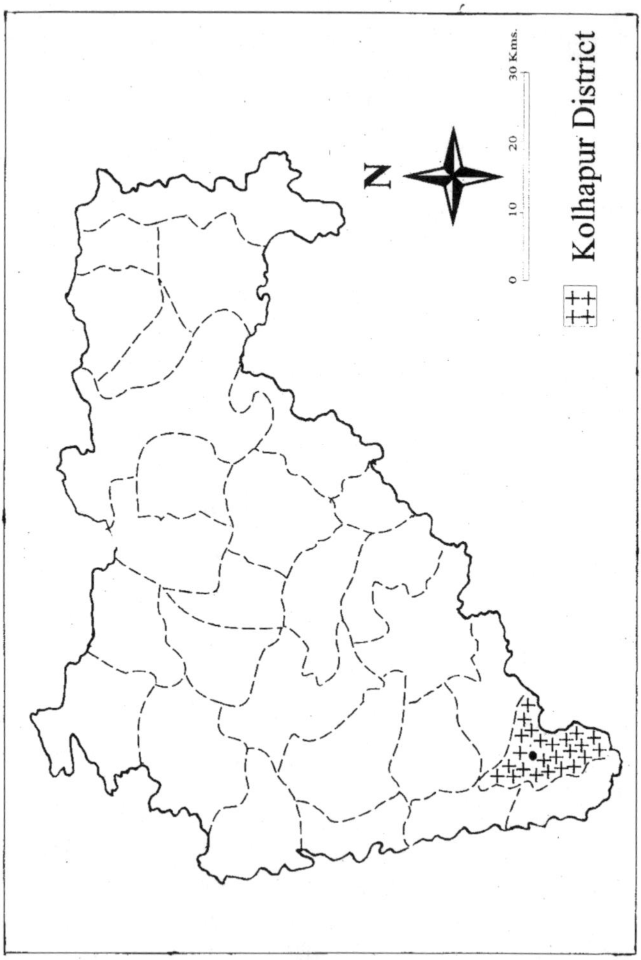

Fig. 1: Map of Maharashtra

predators viz. Lady bird beetle *Menochilus sexmaculatus* Fab., Tiger beetle *Eudema angulatum* Horn. and a praying mantid *Hierodula bipapilla* (Serville -Audinet) as these are abundant and potential biological pest control agents. The text also contains biology and the characteristics of habitats of predators and essential prey species available in the habitat and pest-predator index of agricultural crops (Fig. 1). The present work will find stimulatory and useful to farmers, students, teachers and researchers in the field of ecology and pest management.

Chapter 2
Rearing of Predators

Rearing of insects has given top priority in biological control programme. Any minor change in technique may leads to drastic results both positive and negative. The basic requirements for successful rearing are (1) Food, (2) Protection from enemies, (3) a suitable physical environment and (4) Optimum conditions for reproduction. Insect rearing is an on going process and more efficient methods are continuously being adopted (Sathe & Bhosale, 2001). The application of suitable room techniques and environmental control system greatly facilitate the production of insects in continuous culture. In the present topic, various materials and methods have been reported for rearing the insects. The details of which are mentioned below:

Materials

Glass Cages (Fig. 2 & 3)

Coccinellid beetles, Carabid beetles and Praying mantids have been reared in two types of glass cages (Fig. 2 & 3). Both cages are quadrangular (size 25 × 25 × 30 cm) in shape. Each type consists wooden base and frame with glass walls on three sides. In one cage (Fig. 2) 4th side is provided with muslin sleeve as a door for handling the insects and the another cage (Fig. 3) consists of a glass door . Both types of cages were used for rearing the predator species such as tiger beetles, preying mantids and their preys. To facilitate the rearing technique effectively, a twig of a host plant along with preys and predators were fixed in a circular paper disc over the beaker containing water and kept in these cages.

Rearing Unit-I (Fig. 4)

Rearing unit I consists conical flask, glass chimney, paper disc and prey food plant twig. A glass chimney which measure 23 cm in height, 4.5 cm in diameter at one end, 6.5 cm at the another and 9.5 cm in diameter at the broadest region, the conical flask (500 ml) which measures 17 cm in height and a circular paper disc with central hole. For the rearing of predators, conical flask was filled with water and a twig of a host plant along with preys was fixed through the hole of circular paper disc so as to reach the basal portion of the twig into the water (Fig. 4). Later, in the unit the twig was covered with glass chimney keeping narrow end over the conical flask, the another end of the chimney was closed by muslin cloth. After setting unit, predators were released into the unit for rearing.

Rearing Unit-II (Fig. 5)

This unit contains plastic container of size 4.5 cm in diameter and 6.5 cm in height, empty electric glass bulb and a circular paper disc. A host plant twig was fitted to the paper disc and placed inside the empty glass bulb as shown in Figure 5. The twig was covered by bulb kept over the plastic container so as to reach the plant twig into the water in the container. For the aeration numerous holes were prepared on the upper part of the container. Rearing unit II has a special feature that it maintains the moisture in the unit which is helpful for increasing longevity of prey food plant material.

Glass trough (Fig. 6)

Glass troughs of 11 cm height and 19 cm diameter were used for rearing the coccinellid beetles, tiger beetles and preying mantids. During the rearing the troughs were covered by muslin cloth.

Petridishes (Fig. 7 & 11a)

Two types of petridishes of 12 cm and 10 cm in diameter and 2.3 cm and 2 cm in height were used for rearing the predator and prey species and for observing behavioural aspects like mating, oviposition, etc.

Plastic containers (Fig. 8)

Small four types of plastic container with 5.5 cm, 4.5 cm, 4.0 cm and 3.2 cm in diameter and 7.5 cm, 5.0 cm, 4.5 cm and 3.7 cm in height respectively were used for handling and rearing the predator/prey species.

Plate I

Fig. 2: Glass Cage with Sleeve, Fig. 3: Glass Cage, Fig. 4: Rearing Unit I, Fig. 5: Rearing Unit II, Fig. 6: Glass Trough, Fig. 7: Petridishes.

In addition, large three types of plastic containers with 9 cm, 8 cm and 7.5 cm in diameter and 17 cm, 15 cm and 16 cm in height respectively were used for rearing the predator / prey species.

Test Tubes (Fig. 9)

Test tubes were used for handling the predators (size 19×2.5, 15×2, 14.5×2.8 cm) in the laboratory.

Specimen Tubes (Fig. 10)

Specimen tubes of size 1.7 cm in diameter and 5.2 cm in height were also used for handling the insects and preserving various stages of life cycle of the beetles, specially coccinellids and its parasitoids .

Insect Boxes (Fig. 11)

Insect boxes (Vaman D. Purohit & Sons made) size 42 × 28 × 6.5 cm were used for keeping the preserved insects.

Methods

Rearing of Coccinellids

From different fields of agricultural crops of Kolhapur district (Fig. 12) adults, pupae and larvae of coccinellids were collected and reared in rearing unit I (Fig. 4) and rearing Unit II (Fig. 5). Newly emerged adults from the field collected material (pupae), males and females allowed to mate in petridishes / glass cages (in pairs). Later, the mated females were introduced into the rearing unit I and II for egg laying and incubation was observed. Newly emerged larvae were fed with natural food preys such as aphids (*Rhophalosiphum maidis* Fitch.) and *Aphis gossipii* (Clover) and allow to grow into four instars, pupae and adults. During the rearing of the predator species, their food material (aphids) were changed by 48 hrs interval. The aphids were provided to the predators by taking them on the host plant twig as described above.

Rearing of praying mantids

Praying mantids were collected from the different agricultural fields of Kolhapur district and reared in rearing glass cages (Fig. 2 & 3) by providing natural preys such as grasshoppers (*Hieroglyphus banian* Fab.), mealy bugs (*Phenacoccus hirsutus* Cr.), and *Spodoptera litura* Fab. caterpillars along with twigs of prey food plants.

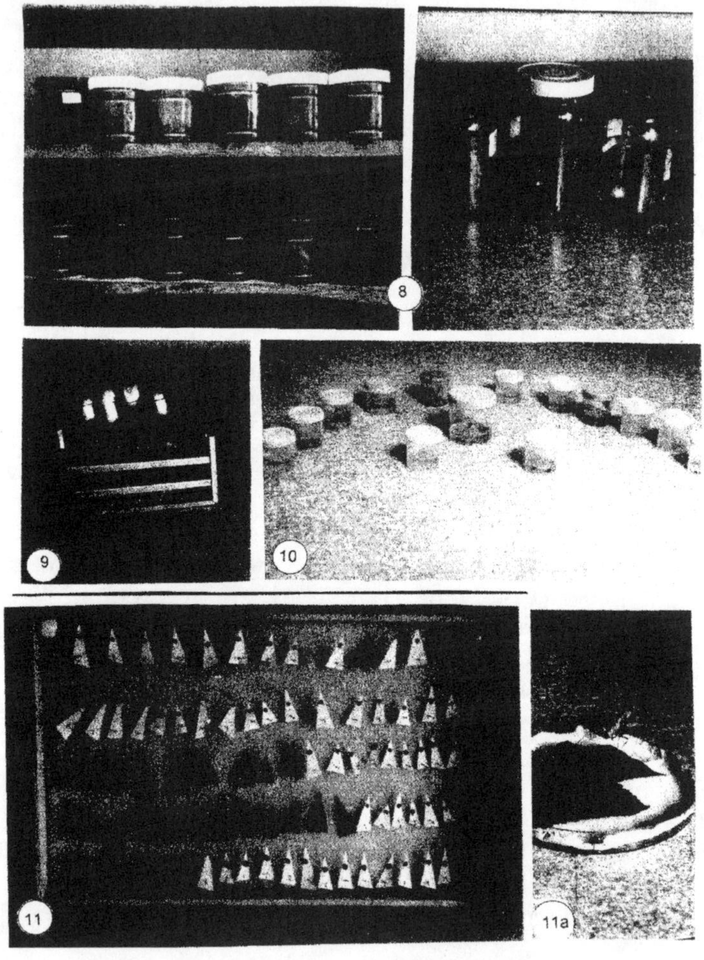

Plate II

Fig. 8: Plastic Containers, Fig. 9: Test Tubes, Fig. 10: Specimen Tubes, Fig. 11: Insect Boxes, Fig. 11a: Nymph of Praying Mantid on Petridish.

Rearing of Tiger beetles

Tiger beetles were collected from the different agricultural fields/ open habitat of Kolhapur district and reared in glass cages (Figs. 2 & 3). The beetles were provided with their natural food specially grasshoppers (*H. banian*), aphids (*A. gossipii* and small caterpillars of *S. litura* along with twigs of prey food plants. For praying mantids and tiger beetles rearing was adopted in glass cages (Fig. 2 & 3) as described in coccinellids.

Chapter 3
Survey and Surveillance

Insect population sampling specially the sampling of insect pests is one of the most useful task in pest management. Sampling of insect is dependent on specialized habits and habitats of the insects. Moreover, sampling of insects with specialized habits such as predators, requires basic knowledge of biology of preys and predators being sampled and techniques often have to be devised for each situation. The predators are usually specific for one or closely related group of insect species. During the larval and adult period predators are restricted to and dependant on an individual of the prey species and thus their reproductive rates are also dependant upon the availability of these preys. In other words, the sampling of predators is the sampling of the predator's prey.

For studying life table statistics the survey of species and successive estimates of the number of insects per unit of land area are necessary. Nearly all other studies of population dynamics of field condition needs survey and surveillance. In general, the survey and surveillance of preys and predators have an immense value in the pest management strategies and will add to knowledge of predator behaviour in relation to the reproduction, population and occurrence of the pest and the crops in the region. For execution and efficient management of pest, study of population dynamics in relation to ecological factors is essential. The validity of such studies is area specific and no such work have so far been reported in and around Kolhapur district. Hence, the present study was made on the predators, *Menochilus sexmaculatus* (Fabricious), Tiger beetle, *Eudema angulatum* Horn and praying mantid *Hierodula bipapilla* S-A. The above three predators are potential biocontrol agents of several agricultural pests in Kolhapur district. Kolhapur is situated in Western Maharashtra (Fig. 1) and is characterized by plain and hilly regions. It has variable rainfall ranging from 700 to 5000 mm with variable intensity (Figs. 13 & 14). It shows various types of agro-ecosystems, forest ecosystems and aquatic ecosystems. The present work will be useful for biological pest control programme of insect pests and understanding the ecology of predators.

Survey of *M. sexmaculatus*, tiger beetle *E. angulatum* (Fig. 25) and praying mantid *H. bipapilla* (Fig. 26) was conducted at four Tahasils (Fig. 12) Hatkanangale, Kagal, Kolhapur & Radhanagari of Kolhapur district. Weekly collections of predator larvae/pupae/adults on jowar/mulberry were made in 100 sq.ft. area for 1 hour in each study centre (Fig. 12). The sample area was changed in successive collections within the same field. Observations were continued in the same field till the harvest of the crop. The collected predators (larvae/pupae/adults) were reared/maintained in the laboratory (25±2°C, R.H. 60-65% 12 hrs photoperiod) on their respective prey species, aphids (*Rhophalosiphum maidis* (Fitch) for *M. sexmaculatus*, caterpillars of *Spodoptera litura* (Fab.) for tiger beetle and praying mantids. During the developmental time of immature stages, observations were also made on the possibilities of parasitoid emergence from larval/pupal forms of predators and on other mortality factors. The per cent parasitism calculated by number of hosts collected and number of parasitoids emerged from them. The mortality caused by other factors was calculated by counting the dead larvae and pupae of the predators during the course of study. The meteorological data mainly temperature, relative humidity and rainfall were considered during the course of studies (years 2000-2002) which is represented in figures 15 to 23. The survey study was restricted on the crop jowar (Fig. 28 to 31) to *M. sexmaculatus* and *E. angulatum* and mulberry to *H. bipapilla*. On sugarcane (Fig. 33) random survey was made by frequent visits.

Survey of *M. sexmaculatus* (Fig. 24)

The field survey of *M. sexmaculatus* on jowar crop at Hatkanangale (Fig. 23), Kagal, Kolhapur (Fig. 29) and Radhanagari (Fig. 31) showed that the appearance of the species were started from 2nd week of July, it reached its peak in the 2nd week of September

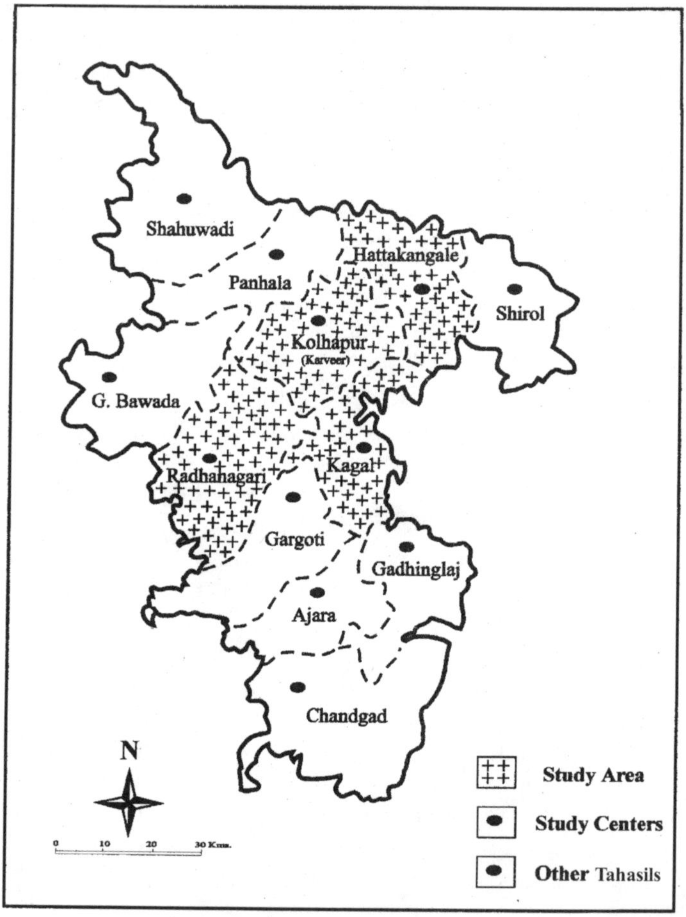

Fig. 12: Map of Kolhapur District Showing Study Area.

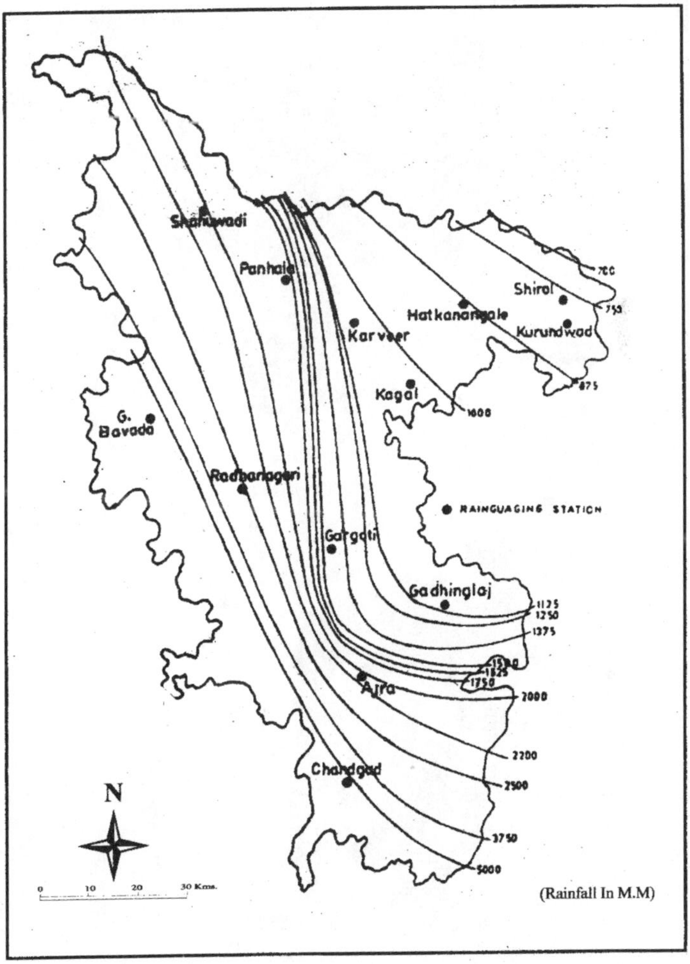

Fig. 13: Average Annual Rainfall of Kolhapur District

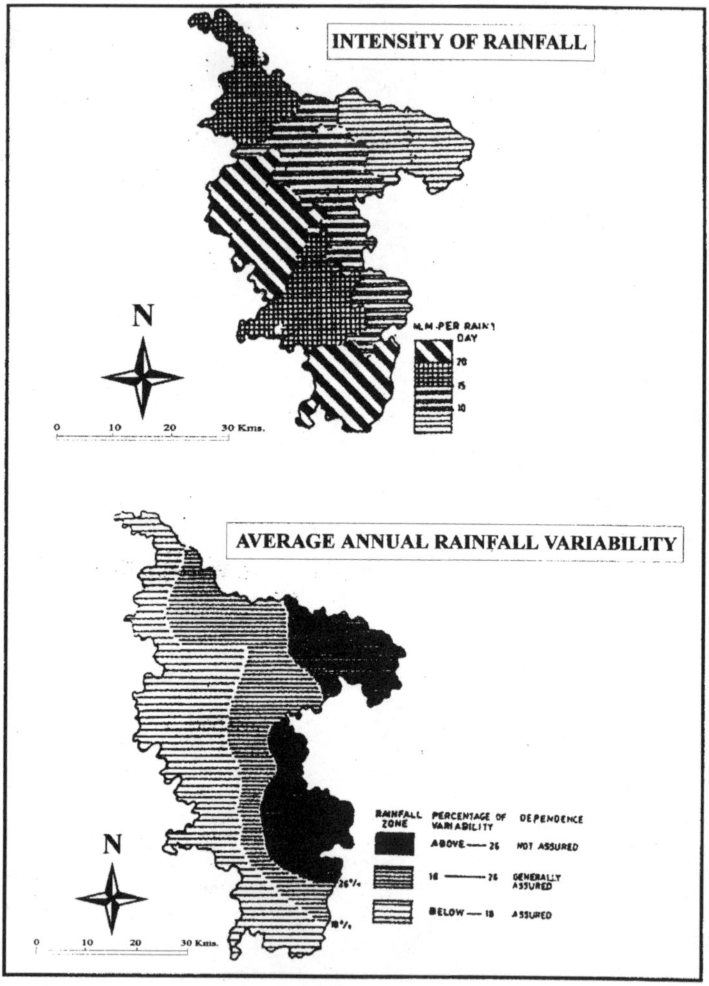

Fig. 14: Map of Kolhapur District Showing Rainfall Intensity and Variability

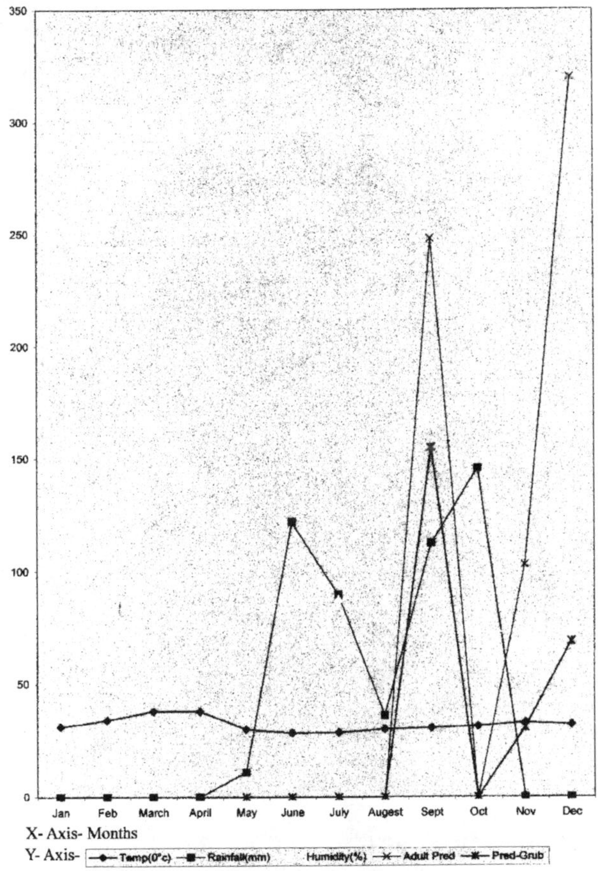

Fig. 15: Seasonal Abundance of *Menochilus sexmaculatus* During 2000 at Kolhapur District

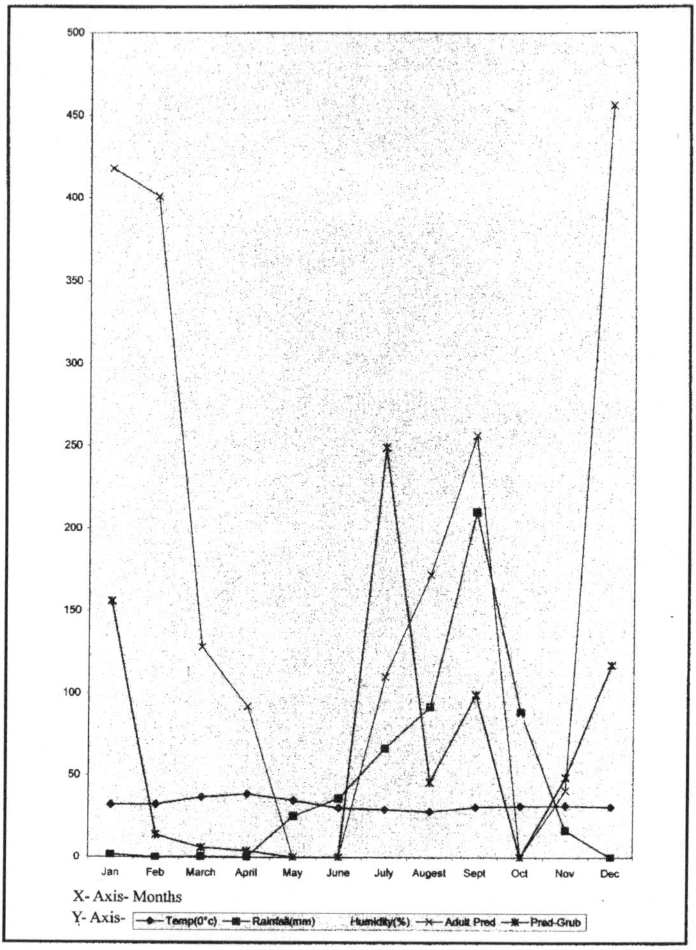

Fig. 16: Seasonal Abundance of *Menochilus sexmaculatus* During 2001 at Kolhapur District

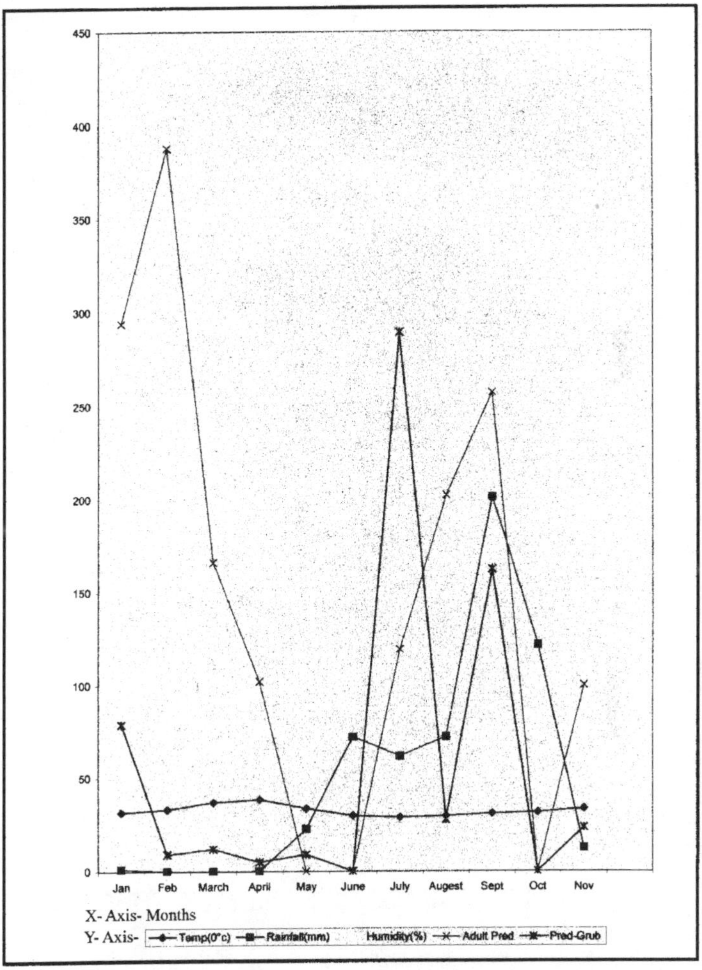

Fig. 17: Seasonal Abundance of *Menochilus sexmaculatus* During 2002 at Kolhapur District

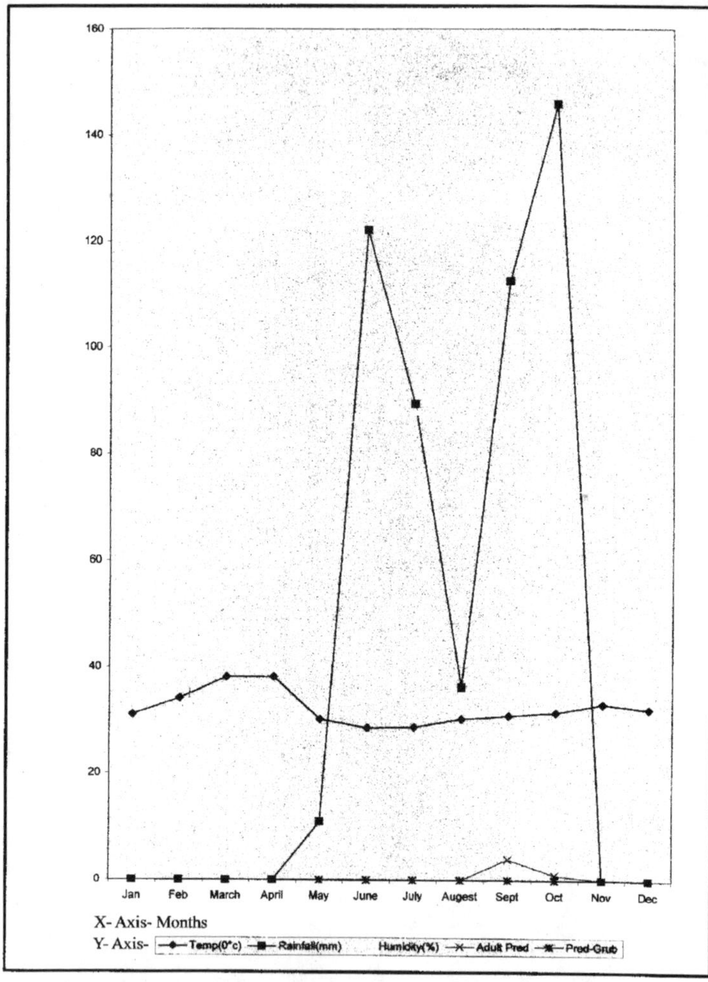

Fig. 18: Seasonal Abundance of *Eudema angulatum* (*Tiger beetle*) During 2000 at Kolhapur District

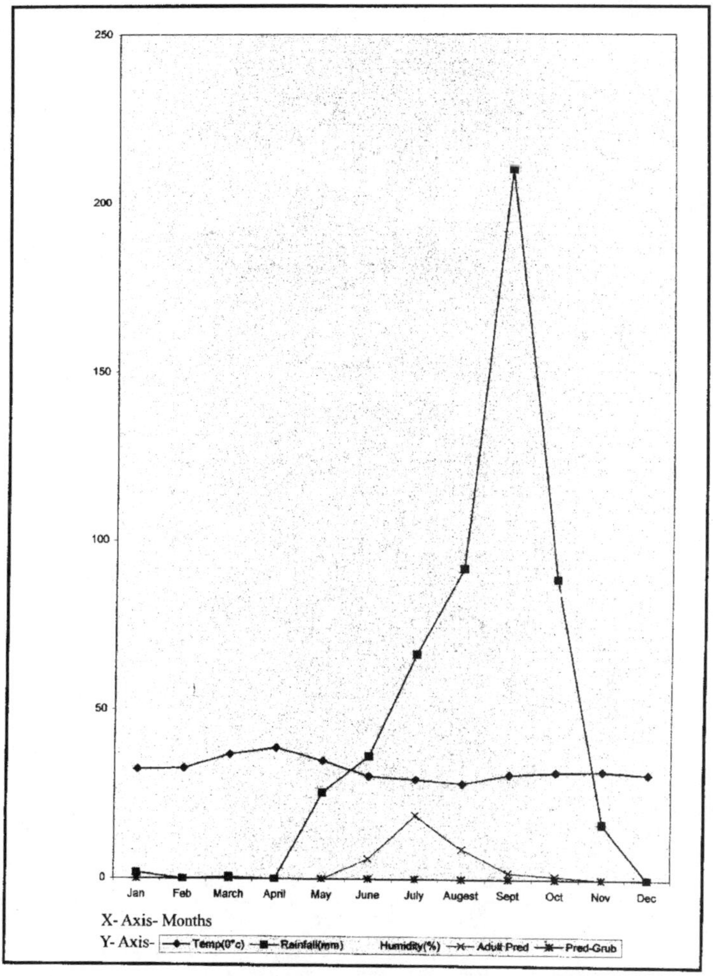

Fig. 19: Seasonal Abundance of *Eudema angulatum* (*Tiger beetle*) During 2001 at Kolhapur District

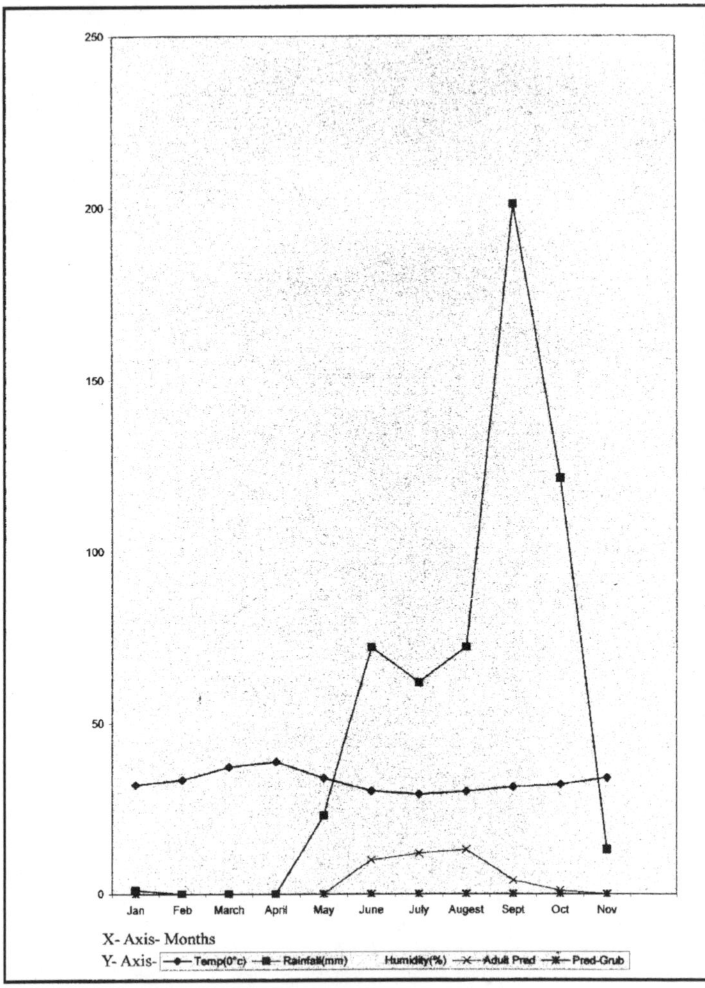

Fig. 20: Seasonal Abundance of *Eudema angulatum* (*Tiger beetle*) During 2002 at Kolhapur District

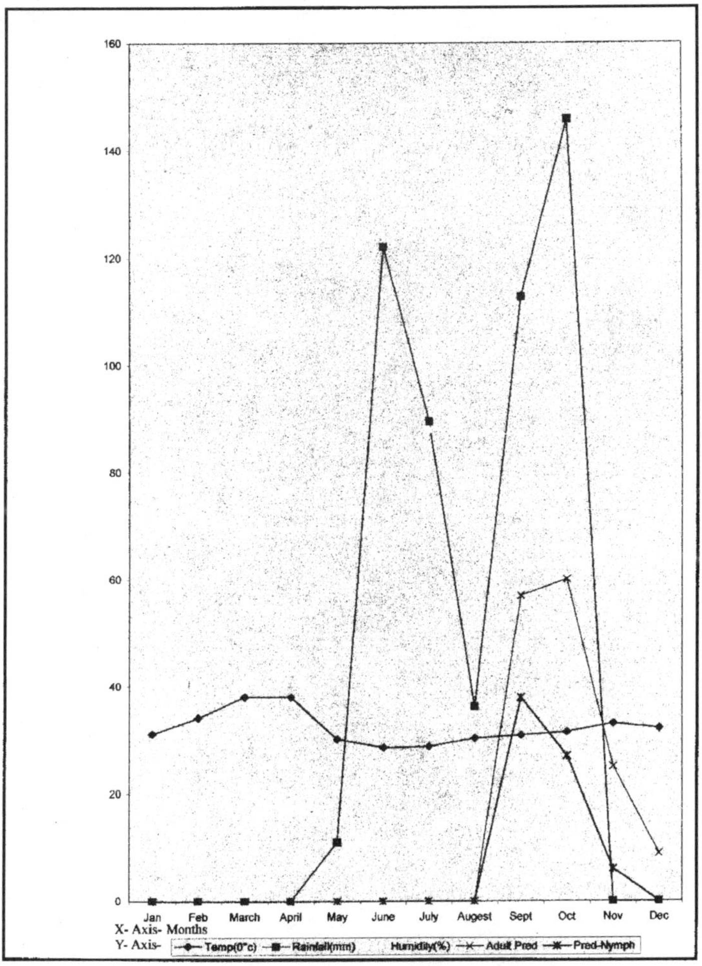

Fig. 21: Seasonal Abundance of *Hierodula bipapilla* (*Praying mantid*) During 2000 at Kolhapur District

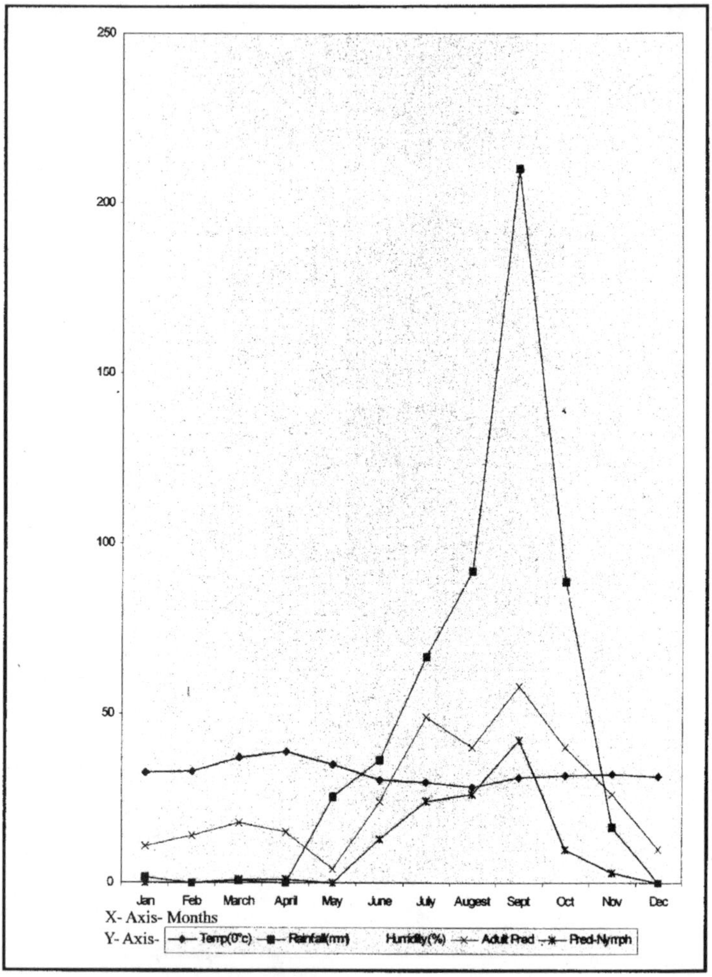

Fig. 22: Seasonal Abundance of *Hierodula bipapilla* (*Praying mantid*) During 2001 at Kolhapur District

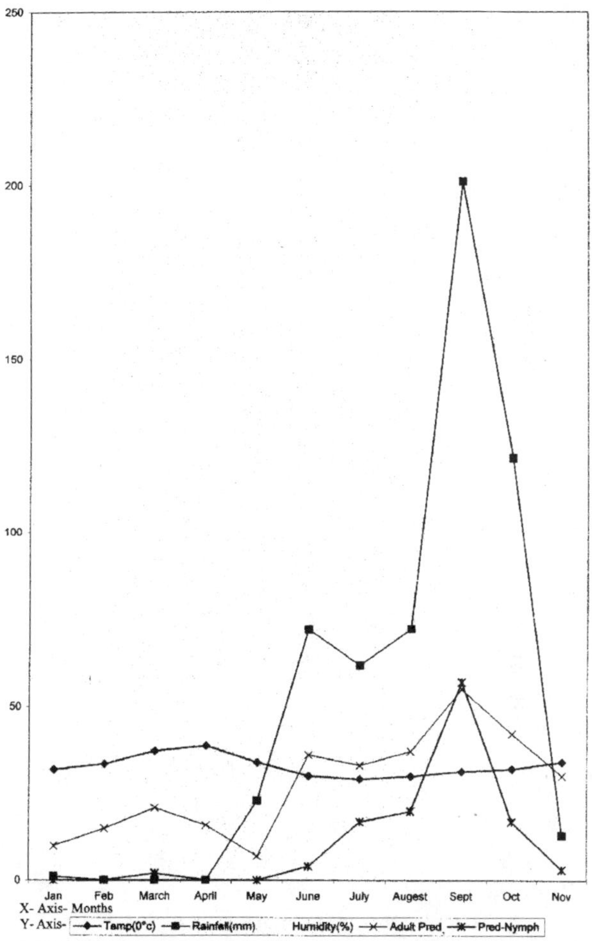

Fig. 23: Seasonal Abundance of *Hierodula bipapilla* (*Praying mantid*) During 2002 at Kolhapur District

during Kharif season while it appeared in the 3rd week of November and reached its peak in the 1st week of February during Rabi season. The seasonal abundance of the species is represented in Tables 1 to 4 and Figures 15 to 17. *M. sexmaculatus* feed on Jowar aphid *R. maidis* on Jowar, Sugarcane aphid *Aphis sachari* Zhnt. on sugarcane, *R. maidis* on maize and Psyllids etc. *M. sexmaculatus* population was found in Radhanagari center on sugarcane during November and December and peaked in January and then it found declined.

Survey of tiger beetle (*E. angulatum*) (Fig. 25)

The field survey of tiger beetles carried out in four centers of present study showed that the tiger beetles found in grassland ecosystem and open ecosystem. They were not found on Jowar, Mulberry (Figs. 30, 32)/Cabbage crops. In general, tiger beetles were common through out the months June, July and August and very few beetles have been observed in September. The results are recorded in Tables 5 to 8. The results indicate that the beetles were most common in June and July (Figs. 18 to 20). It was observed that the beetles found ovipositing as early as May fifteen and late as July fifteen. By mid September hardly one or two beetles could be found. It was also observed that the tiger beetles found rushing in human dwellings at night. These beetles were nocturnal and diurnal in habit. It was noted that tiger beetles had no natural enemies in the form of parasitoids in larval, pupal and adult stages.

Tiger beetles feed on small grasshoppers *Hieroglyphus banian* Fab., *Spodoptera* caterpillars, cutworms (*Agrotis* sp.) and many other small caterpillars, Jassids, aphids, crickets, psyllids etc.

Survey of Praying Mantid (*H. bipapilla*) (Fig. 26)

Survey of praying mantid have been carried out in four centers of Kolhapur district. It was observed that praying mantid peaked in its population in September in all four centers (Tables 9 to 11) (Figs. 21 to 23) and also noted that Oothecae (Figs. 27 & 93) have been parasitized by chalcid parasitoid with 12.5%. Praying mantid has a very wide prey range. It feeds on small insect pests like aphids (Fig. 97 & 100), jassids, mealy bugs, scale insects (Fig. 86), crickets, small caterpillars (Figs. 94 to 96), grasshoppers (Fig. 84), small moths like *Exelastis atomosa* Walsingham and leaf miner moths, psyllids, larvae

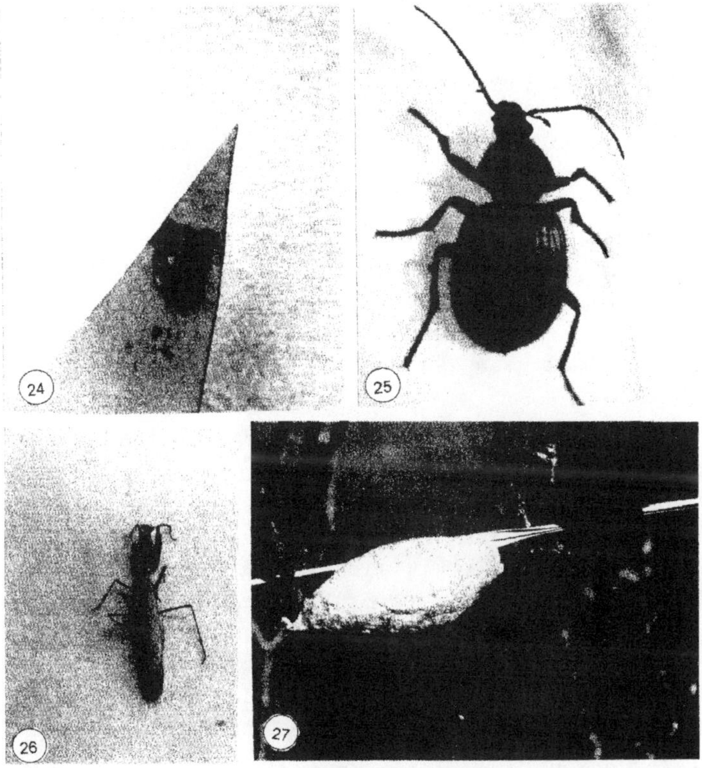

Plate III

Fig. 24: ***Menochilus sexmaculatus*** **(Fab.), Fig. 25:** ***Eudema angulatum*** **(Horn.), Fig. 26:** ***Hierodula bipapilla*** **(S.A.), Fig. 27: Oothica of Praying Mantid.**

of *Erias fabia* and *E. insulana* etc. It also feed on household insect pests such as mosquitoes (*Culex* sp.), cockroaches (*Periplanata americana* L.) (Fig. 98), silver fishes (*Lepisma saccharina* L.) etc.

The average seasonal abundance in Kolhapur district of three predator species viz. *M. sexmaculatus, E. angulatum* and *H. bipapilla* have been represented in Figures 15 to 23.

M. sexmaculatus is potential predator of various homopterous pests such as aphids, *R. maidis, Aphis craccivora* Koch, *Dactynotus compositae* Theobald, *Aphis gossypii* Glover, *Myzus persicae* Sulzer, Psyllids, *Heteropsylla cubana* Crawford, etc. in agro-ecosystem. During the Kharif season, *M. sexmaculatus* attacked *R. maidis* on hybrid variety (CSH-5) of jowar and maize; *A. craccivora* on groundnut, *A. gossypii* on cotton, *A. craccivora* on sunflower and cowpea, *M. persicae* on palak and gawar, etc. While during Rabi season, *M. sexmaculatus* and *Illeis durbarii* S. & B. attacked *R. maidis* on jowar, (*Sorghum vulgare* Pers.), *A. craccivora* on sunflower (*Helianthus annus* Linn.), *M. persicae* on palak (*Spinacia oleracea* Linn.) and gawar (*Cyamposis tetragonolobus* L.), etc.

The infestations of aphids has been recorded in the fields of sorghum during the monsoon (July-September) but not in pre winter (October-November). However, the aphids appeared on the Sorghum in 2nd week of July. The predators appeared on the Sorghum in 2nd week of July and 2nd week of November and reached its peak in 2nd week of September and 1st week of February. The increase in the population of predator might be due to abundance of their food. On the maize crop same interaction of prey and predator was observed. On the groundnut, *A. craccivora* appeared in 1st week of August. Soon after the emergence of pest, the predators appeared and reached its peak in the 4th week of August.

The seasonal abundance of psyllids and its predators was studied by Nakahara *et al.* (1987). They reported that the initial population of psyllids, *H. cubana* was seen during the 3rd week of August following the next week rush on the subabul, *Luceana retusa* Brew. During March the population of psyllid increased like the predators. The population of psyllid gradually declined and reached to the lowest level during the 1st week of May. Similarly, predator adults also gradually declined by 2nd week of May. However, the grub disappeared by the last week of April. These observations are in agreement with those of Diraviam and Virakatmath (1991). In the present study decline in the population of predators was seen from 3rd week of September on hybrid jowar and on the Shalu jowar from 3rd week of February because of the insufficient food. The maturation of host plants lead the aphids to migrate elsewhere.

The present survey study indicated the meteorological factors influence the population dynamics of the prey and predator species. The population of *M. sexmaculatus* on hybrid jowar (CSI I5) reduced in the 3rd week of July 2000 as the rainfall was more, the population also reduced in the month of August 2000, as there was rainfall in all weeks of August 2000, in the September 2000 again, the population increased and then reduced and was almost nil in the month of October due to harvestation of crop.

In the year 2001, *M. sexmaculatus* appeared on the hybrid jowar in 2nd week of July and reached its peak in last two weeks of September. Its population decreased in the 5th of July as there was heavy rainfall.

Seasonal abundance of *M. sexmaculatus*, *I darbari* and *C. yamini S.G.B.* was studied in Maharashtra by Sathe (2002) on sorghum, mulberry, groundnut, vegetable and teak ecosystem. It is observed that *M. sexmaculatus* appears in November and reached its peak in February; *I. darbari* appears on this crop in 2nd week of January and peaks in 3rd week of February. *C. yamini* was not associated with sorghum ecosystem. In mulberry fields, *M. sexmaculatus* reached maximum in its population during January and June. Likely, the population of *I. darbari* reached maximum in July to November (Bhosale, 1994). However, *C. yamini* was not found in mulberry ecosystem. The prey hosts for coccinellids on mulberry were *A. gossypii* and on sorghum *R. maids*.

In groundnut ecosystem, *M. sexmaculatus* and *C. yamini* were dominant, but *I. darbari* never seen. The population of *C. yamini* peaked in August and later in March on groundnut crops of rainfed and irrigated respectively.

In vegetable *Coccinella tranversalis* (Fab.) and *Harmonia octomaculata* (Fab) were commonly found in Maharashtra. *M. persicae* and *Brevicorne brassicae* Linn. were the two aphid prey hosts for above species. The population of *M. sexmaculatus* was found fluctuating on vegetable crops probably due to disliking of the preys. However, in the fields of Miraj *M. sexmaculatus* population was not common either on vegetable, mulberry, groundnut, teak or sorghum during April to June. But, very interestingly for some years *M. sexmaculatus* appeared in Kolhapur on polygoneaceae during April and May (Sathe, 2002).

In teak ecosystem, four species of coccinellids viz., *M. sexmaculatus, Cryptolaemus motrouzieri* Mulsant, *Scymnus (P) coccivora* Diyar and *Jauravia soror* Mots. have been reported from Maharashtra. The preys of above predators were mainly *Icerya purchasi* Maskell, *A. gossypi*, and *Ferrisia virgata* Cockerall. However, *M. sexmaculatus* was not acting as dominant predator. In fact, *C. motrouzieri* played very important role in suppressing *I. purchasi* populations.

Some species of coccinellids displayed close synchrony with their prey so that they reproduce when the prey is increasing, and became quiescent or hibernate when their prey is also less active or dormant. Some aphid finding coccinellids of hot temperature zones showed partial synchrony with their preys. Coccinellids exhibited their greatest activity during the spring but, in summer many coccinellids disappeared even through some aphid populations were available. *C. yamini* disappeared from April to July (Bhosale, 1994).

Many times the small populations of *M. sexmaculatus* and *I. darbari* observed on congress grass and mulberry respectively.

M. sexmaculatus is most abundant and commonest predator of aphids and active almost throughout the year (Omkar & Bind, 1993). The predator some times presents on the crop at the very beginning of the infestation and interact with aphid populations by suppressing it below the level of economic damage level.

M. sexmaculatus has relation with weather factor and population fluctuation in the country. This aspect is studied by several workers in India (Omkar & Bind, 1993; Bhosale, 1994; Sathe & Bhosale, 2001, etc.).

The population of *M. sexmaculatus* reached at its maximum level in mulberry fields during January and June and in guava orchard and teak plantation in July in Karnataka (Bhaskar, 1992). However, in Rajasthan *M. sexmaculatus* population reached in its peak during February and then declines when it predates *Uroleucon compositae* Theobald (Upadhyay *et. al*, 1983). The adult population of *C. coeruleus* increased during January and reached in its peak in June and subsequently declined by October (Nakahara *et. al*, 1987).

In *M. sexmaculatus* only population found peaking during April in summer and September in Kharif season (Shrikanth & Lakkundi, 1990). There exists a highly significant positive correlation between aphid and predator populations in the two seasons. Bhaskar (1992) also reported a significant positive correlations between *M. sexmaculatus* and its prey. Thus, it can be concluded that the population of *M. sexmaculatus* may move from one ecosystem to other in search of prey and or go in diapouse in adverse conditions (temperature, humidity, food etc.) which accounts for the contradictory findings of the present study and there by other authors. The present study also indicates that *E. angulatum* peaked their population in July and August while, *H. bipapilla* peaked in September.

The grubs of *M. sexmaculatus* were found parasitised by Dipterous and hymenopterous parasitoids. The parasitism on grubs were increased to high level as the season progressed specially in winter (November-December). It seems that the high temperature and low or no rainfall provides favourable conditions for a increase of the population in the parasitoids. These results are in agreement with Bhosale (1994). Sathe (1986) studied three hymenopterous parasitoids viz. *Diadegma trichoptilus* Cemeron, *Cotesia orientails* C & N and *C. diurnii* R & N. With respect to field recorded 40% parasitism by these parasitoids in *E. atomosa*. He further reported that *C. orientalis* appeared soon, a week or two and *C. diurnii* slightly later about 4 weeks after the appearance of the pest on crop. In the present study the parasitoid appeared in November, a week after the appearance of predators (grubs). Only *M. sexmaculatus* found parasitized (0.68%) by a chalcid fly (Hymenoptera) and a tachinid fly (2%). The oothicae of praying mantids also found parasitized by hymenopterous chalcid parasitoids upto 12.5%. In India, survey of insect pest predators is badly needed and the study will add great relevance in designing biological pest control programme which is ecofriendly.

Plate IV

Fig. 28: Jowar (CSH-9) Field (Hatkanangale), Fig. 29: Jowar (Shalu) Field (Kolhapur), Fig. 30: Mulberry Field (Kolhapur), Fig. 31: Jowar (Shalu) Field (Radhanagari), Fig. 32: Mulberry Field (Kagal), Fig. 33: Sugarcane Field (Kolhapur).

Plate V

Fig. 34: Mixed Ecosystem (Bamboo, Teak, Sugarcane and Castor) (Kagal), Fig. 35: Teak Ecosystem (Kagal), Fig. 36: Gallardia Ecosystem (Kagal), Fig. 37: Bamboo (Habitat) (Kagal), Fig. 38: Soyabean Agro-ecosystem with Teak Plants (Kagal), Fig. 39: Sugarcane Agro-ecosystem (Kagal).

Plate VI

Fig. 40: Jowar Agro-ecosystem (Kagal), Fig. 41: Mixed agro-ecosystem (Sugarcane+Sunhemp) (Kagal), Fig. 42: Cotton Leaf with *Coccinella* Predator (Kagal), Fig. 43: Tobacco Leaf with *Coccinella* Predator (Kagal), Fig. 44: Nymph of Praying Mantid on Soyabean (Kagal), Fig. 45: Cabbage.

Plate VII

Fig. 46: Jowar agro-ecosystem (Hatkanangale), Fig. 47: Cabbage agro-ecosystem (Hatkanangale), Fig. 48: Brinjal agro-ecosystem (Hatkanangale), Fig. 49: Bamboo Ecosystem (Hatkanangale), Fig. 50: Tomato agro-ecosystem (Hatkanangale), Fig. 51: Chilli agro-ecosystem (Hatkanangale).

Table 1: Collection data of *M. sexmaculatus* on hybrid Jowar (CSH-9) for 2000-2001 at Hatkanangale

Month	*Date of Collection*	*Adult Predators*	*Predator Grubs*	*Parasitoids Emerged (Larval)*	*Unknown Mortality (%)*
September	03-09-2000	78	113	-	-
	17-09-2000	210	56	-	-
October	01-10-2000	-	-	-	-
	15-10-2000	-	-	-	-
	29-10-2000	-	-	-	-
November	12-11-2000	0	-	-	-
	26-11-2000	110	35	-	3
December	10-12-2000	155	35	-	3
	24-12-2000	175	38	3	5
January	10-01-2001	203	14	-	6
	24-10-2001	222	40	-	-
February	04-02-2001	238	17	-	-
	28-02-2001	173	-	-	-
March	04-03-2001	75	8	-	-
	18-03-2001	63	-	-	-
April	08-04-2001	54	4	-	-
	22-04-2001	49	-	-	-
May	06-05-2001	-	-	-	-
	20-05-2001	-	-	-	-
June	03-06-2001	-	-	-	-
	17-06-2001	-	-	-	-
July	01-07-2001	-	113	-	-
	15-07-2001	12	126	-	7
	29-07-2001	98	21	-	-
August	12-08-2001	70	21	-	-
	28-08-2001	96	8	-	2
September	09-09-2001	65	100	-	-
	23-09-2001	196	-	-	-

Contd...

Table 1–Contd...

Month	*Date of Collection*	*Adult Predators*	*Predator Grubs*	*Parasitoids Emerged (Larval)*	*Unknown Mortality (%)*
October	07-10-2001	-	-	-	-
	21-10-2001	-	-	-	-
November	04-11-2001	-	-	-	-
	18-11-2001	40	7	5	-
December	02-12-2001	135	7	-	-
	16-12-2001	165	65	5	-
	30-12-2001	170	30	-	-
January	13-01-2002	190	52	-	-
	27-01-2002	178	35	-	4
February	10-02-2002	215	10	-	-
	24-02-2002	170	-	-	-
March	10-03-2002	102	10	-	-
	24-03-2002	89	-	-	-
April	07-04-2002	56	3	-	-
	21-04-2002	45	-	-	-
May	05-05-2002	-	-	-	-
	19-05-2002	-	-	-	-
June	02-06-2002	-	-	-	-
	16-06-2002	-	-	-	-
	30-06-2002	-	-	-	-
July	14-07-2002	10	125	-	-
	28-07-2002	100	150	-	-
August	11-08-2002	80	16	-	-
	25-08-2002	105	7	-	-
September	08-09-2002	77	114	-	-
	22-09-2002	215	50	-	-
October	06-10-2002	-	-	-	-
	20-10-2002	-	-	-	-
November	03-11-2002	0	-	-	-
	17-11-2002	105	25	-	2

Table 2: Collection data of *M. sexmaculatus* on hybrid Jowar (CSH-9) for 2000-2001 at Kagal

Month	*Date of Collection*	*Adult Predators*	*Predator Grubs*	*Parasitoids Emerged (Larval)*	*Unknown Mortality (%)*
September	03-09-2000	70	103	-	-
	17-09-2000	200	54	-	-
October	01-10-2000	-	-	-	-
	15-10-2000	-	-	-	-
	29-10-2000	-	-	-	-
November	12-11-2000	0	-	-	-
	26-11-2000	99	32	-	2
December	10-12-2000	150	35	-	3
	24-12-2000	170	38	4	-
January	10-01-2001	206	142	-	5
	24-01-2001	211	38	-	-
February	04-02-2001	224	15	-	-
	18-02-2001	175	-	-	-
March	04-03-2001	65	7	-	-
	18-03-2001	60	-	-	-
April	08-04-2001	45	3	-	-
	22-04-2001	40	-	-	-
May	06-05-2001	-	-	-	-
	20-05-2001	-	-	-	-
June	03-06-2001	-	-	-	-
	17-06-2001	-	-	-	-
July	01-07-2001	-	-	-	-
	15-07-2001	14	125	-	8
	29-07-2001	100	130	-	-
August	12-08-2001	80	22	-	-
	28-08-2001	99	6	-	2
September	09-09-2001	63	99	-	-
	23-09-2001	200	11	-	-
October	07-10-2001	-	-	-	-
	21-10-2001	-	-	-	-

Contd...

Table 2–Contd...

Month	*Date of Collection*	*Adult Predators*	*Predator Grubs*	*Parasitoids Emerged (Larval)*	*Unknown Mortality (%)*
November	04-11-2001	-	-	-	-
	18-11-2001	40	66	3	-
December	02-12-2001	130	55	-	-
	16-12-2001	160	33	-	-
	30-12-2001	180	40	-	-
January	13-01-2002	102	55	-	-
	27-01-2002	175	34	-	-
February	10-02-2002	220	8	-	-
	24-02-2002	172	-	-	-
March	10-03-2002	98	12	-	-
	24-03-2002	70	-	-	-
April	07-04-2002	60	4	-	-
	21-04-2002	49	-	-	-
May	05-05-2002	-	-	-	-
	19-05-2002	-	-	-	-
June	02-06-2002	-	-	-	-
	16-06-2002	-	-	-	-
	30-06-2002	-	-	-	-
July	14-07-2002	12	130	-	-
	28-07-2002	115	160	-	-
August	11-08-2002	85	14	-	-
	25-08-2002	125	10	-	-
September	08-09-2002	79	110	-	-
	22-09-2002	203	60	-	-
October	06-10-2002	-	-	-	-
	20-10-2002	-	-	-	-
November	03-11-2002	0	-	-	-
	17-11-2002	97	23	-	-

Table 3: Collection data of *M. sexmaculatus* on hybrid Jowar (CSH-9) for 2000-2001 at Kolhapur

Month	*Date of Collection*	*Adult Predators*	*Predator Grubs*	*Parasitoids Emerged (Larval)*	*Unknown Mortality (%)*
September	03-09-2000	79	112	-	-
	17-09-2000	208	55	-	-
October	01-10-2000	-	-	-	-
	15-10-2000	-	-	-	-
	29-10-2000	-	-	-	-
November	12-11-2000	0	-	-	-
	26-11-2000	112	33	-	-
December	10-12-2000	156	40	3	6
	24-12-2000	180	45	-	-
January	10-01-2001	210	150	-	7
	24-01-2001	230	60	-	-
February	04-02-2001	240	10	-	-
	18-02-2001	175	-	-	-
March	04-03-2001	80	6	-	-
	18-03-2001	65	-	-	-
April	08-04-2001	59	5	-	-
	22-04-2001	45	-	-	-
May	06-05-2001	-	-	-	-
	20-05-2001	-	-	-	-
June	03-06-2001	-	-	-	-
	17-06-2001	-	-	-	-
July	01-07-2001	-	-	-	-
	15-07-2001	14	120	-	8
	29-07-2001	105	140	-	-
August	12-08-2001	90	29	-	-
	28-08-2001	100	7	-	3
September	09-09-2001	66	102	-	-
	23-09-2001	203	-	-	-
October	07-10-2001	-	-	-	-
	21-10-2001	-	-	-	-

Contd...

Table 3–Contd...

Month	*Date of Collection*	*Adult Predators*	*Predator Grubs*	*Parasitoids Emerged (Larval)*	*Unknown Mortality (%)*
November	04-11-2001	-	-	-	-
	18-11-2001	50	67	3	-
December	02-12-2001	140	59	3	-
	16-12-2001	170	35	-	-
	30-12-2001	135	45	-	-
January	13-01-2002	105	50	-	-
	27-01-2002	180	40	-	-
February	10-02-2002	225	10	-	-
	24-02-2002	180	-	-	-
March	10-03-2002	97	15	-	-
	24-03-2002	75	-	-	-
April	07-04-2002	65	5	-	-
	21-04-2002	50	-	-	-
May	05-05-2002	-	-	-	-
	19-05-2002	-	-	-	-
June	02-06-2002	-	-	-	-
	16-06-2002	-	-	-	-
	30-06-2002	-	-	-	-
July	14-07-2002	10	140	-	-
	28-07-2002	120	170	-	-
August	11-08-2002	95	18	-	-
	25-08-2002	130	20	-	-
September	08-09-2002	83	108	-	-
	22-09-2002	210	70	-	-
October	06-10-2002	-	-	-	-
	20-10-2002	-	-	-	-
November	03-11-2002	-	-	-	-
	17-11-2002	101	25	-	2

Table 4: Collection data of *M. sexmaculatus* on hybrid Jowar (CSH-9) for 2000-2001 at Radhanagari

Month	*Date of Collection*	*Adult Predators*	*Predator Grubs*	*Parasitoids Emerged (Larval)*	*Unknown Mortality (%)*
September	03-09-2000	59	92	-	-
	17-09-2000	88	35	-	-
October	01-10-2000	-	-	-	-
	15-10-2000	-	-	-	-
	29-10-2000	-	-	-	-
November	12-11-2000	-	-	-	-
	26-11-2000	92	23	-	-
December	10-12-2000	136	25	2	4
	24-12-2000	160	20	-	-
January	10-01-2001	180	135	-	5
	24-01-2001	210	45	-	-
February	04-02-2001	230	15	-	-
	28-02-2001	150	-	-	-
March	04-03-2001	62	4	-	-
	18-03-2001	42	-	-	-
April	08-04-2001	39	4	-	-
	22-04-2001	25	-	-	-
May	06-05-2001	-	-	-	-
	20-05-2001	-	-	-	-
June	03-06-2001	-	-	-	-
	17-06-2001	-	-	-	-
July	01-07-2001	-	-	-	-
	15-07-2001	10	100	-	6
	29-07-2001	85	123	-	-
August	12-08-2001	71	6	-	3
	28-08-2001	82	86	-	-
September	09-09-2001	46	86	-	-
	23-09-2001	183	-	-	-
October	07-10-2001	-	-	-	-
	21-10-2001	-	-	-	-

Contd...

Table 4–Contd...

Month	*Date of Collection*	*Adult Predators*	*Predator Grubs*	*Parasitoids Emerged (Larval)*	*Unknown Mortality (%)*
November	04-11-2001	-	-	-	-
	18-11-2001	35	56	2	-
December	02-12-2001	122	51	3	-
	16-12-2001	150	25	-	-
	30-12-2001	165	23	-	-
January	13-01-2002	83	30	-	-
	27-01-2002	163	20	-	-
February	10-02-2002	205	8	-	-
	24-02-2002	164	-	-	-
March	10-03-2002	77	10	-	-
	24-03-2002	55	-	-	-
April	07-04-2002	45	8	-	-
	21-04-2002	38	-	-	-
May	05-05-2002	-	-	-	-
	19-05-2002	-	-	-	-
June	02-06-2002	-	-	-	-
	16-06-2002	-	-	-	-
	30-06-2002	-	-	-	-
July	14-07-2002	8	126	-	-
	28-07-2002	100	150	-	-
August	11-08-2002	80	12	-	-
	25-08-2002	110	15	-	-
September	08-09-2002	63	88	-	-
	22-09-2002	98	50	-	-
October	06-10-2002	-	-	-	-
	20-10-2002	-	-	-	-
November	03-11-2002	8	-	-	-
	17-11-2002	91	22	-	2

Table 5: Collection data of Tiger beetle *E. angulatum* in grassland ecosystem for 2000-2002 at Hatkanangale

Month	*Date of Collection*	*Adult Predators*	*Predator Grubs*	*Parasitoids Emerged (Larval)*	*Unknown Mortality (%)*
September	03-09-2000	3	-	-	-
	17-09-2000	2	-	-	-
October	01-10-2000	-	-	-	-
	15-10-2000	3	-	-	-
	29-10-2000	-	-	-	-
November	12-11-2000	-	-	-	-
	26-11-2000	-	-	-	-
December	10-12-2000	-	-	-	-
	24-12-2000	-	-	-	-
January	10-01-2001	-	-	-	-
	24-01-2001	-	-	-	-
February	04-02-2001	-	-	-	-
	18-02-2001	-	-	-	-
March	04-03-2001	-	-	-	-
	18-03-2001	-	-	-	-
April	08-04-2001	-	-	-	-
	22-04-2001	-	-	-	-
May	06-05-2001	-	-	-	-
	20-05-2001	-	-	-	-
June	03-06-2001	5	-	-	1
	17-06-2001	8	-	-	-
July	01-07-2001	10	-	-	-
	15-07-2001	7	-	-	-
	29-07-2001	2	-	-	-
August	12-08-2001	8	-	-	-
	28-08-2001	3	-	-	-
September	09-09-2001	1	-	-	-
	23-09-2001	2	-	-	-
October	07-10-2001	1	-	-	-
	21-10-2001	-	-	-	-

Contd...

Table 5–Contd...

Month	Date of Collection	Adult Predators	Predator Grubs	Parasitoids Emerged (Larval)	Unknown Mortality (%)
November	04-11-2001	-	-	-	-
	18-11-2001	-	-	-	-
December	02-12-2001	-	-	-	-
	16-12-2001	-	-	-	-
	30-12-2001	-	-	-	-
January	13-01-2002	-	-	-	-
	27-01-2002	-	-	-	-
February	10-02-2002	-	-	-	-
	24-02-2002	-	-	-	-
March	10-03-2002	-	-	-	-
	24-03-2002	-	-	-	-
April	07-04-2002	-	-	-	-
	21-04-2002	-	-	-	-
May	05-05-2002	-	-	-	-
	19-05-2002	-	-	-	-
June	02-06-2002	-	-	-	-
	16-06-2002	2	-	-	-
	30-06-2002	10	-	-	-
July	14-07-2002	2	-	-	-
	28-07-2002	6	-	-	-
August	11-08-2002	10	-	-	1
	25-08-2002	4	-	-	-
September	08-09-2002	2	-	-	-
	22-09-2002	1	-	-	-
October	06-10-2002	-	-	-	-
	20-10-2002	-	-	-	-
November	03-11-2002	-	-	-	-
	17-11-2002	-	-	-	-

Table 6: Collection data of Tiger beetle *E. angulatum* in grassland ecosystem for 2000-2002 at Kagal

Month	*Date of Collection*	*Adult Predators*	*Predator Grubs*	*Parasitoids Emerged (Larval)*	*Unknown Mortality (%)*
September	03-09-2000	2	-	-	-
	17-09-2000	1	-	-	-
October	01-10-2000	-	-	-	-
	15-10-2000	-	-	-	-
	29-10-2000	-	-	-	-
November	12-11-2000	-	-	-	-
	26-11-2000	-	-	-	-
December	10-12-2000	-	-	-	-
	24-12-2000	-	-	-	-
January	10-01-2001	-	-	-	-
	24-01-2001	-	-	-	-
February	04-02-2001	-	-	-	-
	28-02-2001	-	-	-	-
March	04-03-2001	-	-	-	-
	18-03-2001	-	-	-	-
April	08-04-2001	-	-	-	-
	22-04-2001	-	-	-	-
May	06-05-2001	-	-	-	-
	20-05-2001	-	-	-	-
June	03-06-2001	-	-	-	-
	17-06-2001	4	-	-	-
July	01-07-2001	10	-	-	-
	15-07-2001	8	-	-	-
	29-07-2001	6	-	-	-
August	12-08-2001	9	-	-	-
	28-08-2001	2	-	-	1
September	09-09-2001	3	-	-	-
	23-09-2001	1	-	-	-
October	07-10-2001	-	-	-	-
	21-10-2001	-	-	-	-

Contd...

Table 6–Contd...

Month	Date of Collection	Adult Predators	Predator Grubs	Parasitoids Emerged (Larval)	Unknown Mortality (%)
November	04-11-2001	-	-	-	-
	18-11-2001	-	-	-	-
December	02-12-2001	-	-	-	-
	16-12-2001	-	-	-	-
	30-12-2001	-	-	-	-
January	13-01-2002	-	-	-	-
	27-01-2002	-	-	-	-
February	10-02-2002	-	-	-	-
	24-02-2002	-	-	-	-
March	10-03-2002	-	-	-	-
	24-03-2002	-	-	-	-
April	07-04-2002	-	-	-	-
	21-04-2002	-	-	-	-
May	05-05-2002	-	-	-	-
	19-05-2002	-	-	-	-
June	02-06-2002	-	-	-	-
	16-06-2002	4	-	-	-
	30-06-2002	7	-	-	-
July	14-07-2002	9	-	-	-
	28-07-2002	5	-	-	-
August	11-08-2002	10	-	-	1
	25-08-2002	6	-	-	-
September	08-09-2002	4	-	-	-
	22-09-2002	2	-	-	-
October	06-10-2002	1	-	-	-
	20-10-2002	-	-	-	-
November	03-11-2002	-	-	-	-
	17-11-2002	-	-	-	-

Table 7: Collection data of Tiger beetle *E. angulatum* in grassland ecosystem for 2000-2002 at Kolhapur

Month	*Date of Collection*	*Adult Predators*	*Predator Grubs*	*Parasitoids Emerged (Larval)*	*Unknown Mortality (%)*
September	03-09-2000	4	-	-	-
	17-09-2000	1	-	-	-
October	01-10-2000	-	-	-	-
	15-10-2000	2	-	-	-
	29-10-2000	-	-	-	-
November	12-11-2000	-	-	-	-
	26-11-2000	-	-	-	-
December	10-12-2000	-	-	-	-
	24-12-2000	-	-	-	-
January	10-01-2001	-	-	-	-
	24-01-2001	-	-	-	-
February	04-02-2001	-	-	-	-
	18-02-2001	-	-	-	-
March	04-03-2001	-	-	-	-
	18-03-2001	-	-	-	-
April	08-04-2001	-	-	-	-
	22-04-2001	-	-	-	-
May	06-05-2001	-	-	-	-
	20-05-2001	-	-	-	-
June	03-06-2001	-	-	-	-
	17-06-2001	3	-	-	-
July	01-07-2001	7	-	-	-
	15-07-2001	9	-	-	1
	29-07-2001	5	-	-	-
August	12-08-2001	6	-	-	-
	28-08-2001	3	-	-	-
September	09-09-2001	1	-	-	-
	23-09-2001	-	-	-	-
October	07-10-2001	2	-	-	-
	21-10-2001	-	-	-	-

Contd...

Table 7–Contd...

Month	*Date of Collection*	*Adult Predators*	*Predator Grubs*	*Parasitoids Emerged (Larval)*	*Unknown Mortality (%)*
November	04-11-2001	-	-	-	-
	18-11-2001	-	-	-	-
December	02-12-2001	-	-	-	-
	16-12-2001	-	-	-	-
	30-12-2001	-	-	-	-
January	13-01-2002	-	-	-	-
	27-01-2002	-	-	-	-
February	10-02-2002	-	-	-	-
	24-02-2002	-	-	-	-
March	10-03-2002	-	-	-	-
	24-03-2002	-	-	-	-
April	07-04-2002	-	-	-	-
	21-04-2002	-	-	-	-
May	05-05-2002	-	-	-	-
	19-05-2002	-	-	-	-
June	02-06-2002	-	-	-	-
	16-06-2002	2	-	-	-
	30-06-2002	4	-	-	-
July	14-07-2002	8	-	-	-
	28-07-2002	6	-	-	-
August	11-08-2002	10	-	-	1
	25-08-2002	5	-	-	-
September	08-09-2002	2	-	-	-
	22-09-2002	1	-	-	-
October	06-10-2002	-	-	-	-
	20-10-2002	2	-	-	-
November	03-11-2002	-	-	-	-
	17-11-2002	-	-	-	-

Table 8: Collection data of Tiger beetle *E. angulatum* in grassland ecosystem for 2000-2002 at Radhanagari

Month	*Date of Collection*	*Adult Predators*	*Predator Grubs*	*Parasitoids Emerged (Larval)*	*Unknown Mortality (%)*
September	03-09-2000	2	-	-	-
	17-09-2000	1	-	-	-
October	01-10-2000	-	-	-	-
	15-10-2000	-	-	-	-
	29-10-2000	-	-	-	-
November	12-11-2000	-	-	-	-
	26-11-2000	-	-	-	-
December	10-12-2000	-	-	-	-
	24-12-2000	-	-	-	-
January	10-01-2001	-	-	-	-
	24-01-2001	-	-	-	-
February	04-02-2001	-	-	-	-
	28-02-2001	-	-	-	-
March	04-03-2001	-	-	-	-
	18-03-2001	-	-	-	-
April	08-04-2001	-	-	-	-
	22-04-2001	-	-	-	-
May	06-05-2001	-	-	-	-
	20-05-2001	-	-	-	-
June	03-06-2001	-	-	-	-
	17-06-2001	3	-	-	-
July	01-07-2001	5	-	-	-
	15-07-2001	4	-	-	1
	29-07-2001	3	-	-	-
August	12-08-2001	1	-	-	-
	28-08-2001	4	-	-	-
September	09-09-2001	-	-	-	-
	23-09-2001	2	-	-	-
October	07-10-2001	-	-	-	-
	21-10-2001	-	-	-	-

Contd...

Table 8–Contd...

Month	*Date of Collection*	*Adult Predators*	*Predator Grubs*	*Parasitoids Emerged (Larval)*	*Unknown Mortality (%)*
November	04-11-2001	-	-	-	-
	18-11-2001	-	-	-	-
December	02-12-2001	-	-	-	-
	16-12-2001	-	-	-	-
	30-12-2001	-	-	-	-
January	13-01-2002	-	-	-	-
	27-01-2002	-	-	-	-
February	10-02-2002	-	-	-	-
	24-02-2002	-	-	-	-
March	10-03-2002	-	-	-	-
	24-03-2002	-	-	-	-
April	07-04-2002	-	-	-	-
	21-04-2002	-	-	-	-
May	05-05-2002	-	-	-	-
	19-05-2002	-	-	-	-
June	02-06-2002	-	-	-	-
	16-06-2002	4	-	-	-
	30-06-2002	7	-	-	1
July	14-07-2002	5	-	-	-
	28-07-2002	9	-	-	1
August	11-08-2002	2	-	-	-
	25-08-2002	4	-	-	-
September	08-09-2002	2	-	-	-
	22-09-2002	1	-	-	-
October	06-10-2002	-	-	-	-
	20-10-2002	-	-	-	-
November	03-11-2002	-	-	-	-
	17-11-2002	-	-	-	-

Table 9: Collection data of Praying mantid *H. bipapilla* on Mulberry for 2000-2002 at Hatkanangale

Month	*Date of Collection*	*Adult Predators*	*Predator Nymphs*	*No. of Oothicae*	*No. of Oothicae Parasitized*	*Unknown Mortality (%)*
September	03-09-2000	25	20	3	1	1
	17-09-2000	27	7	-	-	-
October	01-10-2000	25	10	1	-	-
	15-10-2000	23	9	-	-	-
	29-10-2000	18	6	-	-	-
November	12-11-2000	15	3	-	-	-
	26-11-2000	11	2	-	-	-
December	10-12-2000	5	-	-	-	-
	24-12-2000	4	-	-	-	-
January	07-01-2001	5	-	-	-	-
	21-01-2001	5	-	-	-	-
February	04-02-2001	7	-	-	-	-
	18-02-2001	8	-	-	-	-
March	04-03-2001	9	-	-	-	-
	18-03-2001	9	1	-	-	-
April	08-04-2001	10	1	-	-	-
	22-04-2001	5	-	-	-	-
May	06-05-2001	2	-	-	-	-
	20-05-2001	3	-	-	-	-
June	03-06-2001	12	1	-	-	-
	17-06-2001	13	-	-	-	-
July	01-07-2001	15	4	-	-	-
	15-07-2001	17	3	1	0	-
	29-07-2001	18	4	-	-	-
August	12-08-2001	20	32	2	1	2
	28-08-2001	21	6	-	-	-
September	09-09-2001	33	25	3	-	1
	23-09-2001	26	7	1	0	-

Contd...

Table 9–Contd...

Month	*Date of Collection*	*Adult Predators*	*Predator Nymphs*	*No. of Oothicae*	*No. of Oothicae Parasitized*	*Unknown Mortality (%)*
October	07-10-2001	24	6	-	-	-
	21-10-2001	20	-	-	-	-
November	04-11-2001	18	2	-	-	-
	18-11-2001	9	-	-	-	-
December	02-12-2001	4	-	-	-	-
	16-12-2001	3	-	-	-	-
	30-12-2001	2	-	-	-	-
January	13-01-2002	5	-	-	-	-
	27-01-2002	6	-	-	-	-
February	10-02-2002	7	-	-	-	-
	24-02-2002	9	-	-	-	-
March	16-03-2002	10	-	-	-	-
	24-03-2002	10	2	-	-	-
April	07-04-2002	10	1	-	-	-
	21-04-2002	6	-	-	-	-
May	05-05-2002	3	-	-	-	-
	19-05-2002	4	-	-	-	-
June	02-06-2002	10	2	-	-	-
	16-06-2002	12	4	-	-	-
	30-06-2002	14	3	-	-	-
July	14-07-2002	16	7	2	0	1
	28-07-2002	18	3	-	-	-
August	11-08-2002	17	5	-	-	-
	25-08-2002	20	4	1	-	-
September	09-09-2002	23	18	2	0	1
	22-09-2002	30	26	4	2	2
October	06-10-2002	20	12	-	-	-
	20-10-2002	21	7	-	-	-
November	03-11-2002	18	4	-	-	-
	17-11-2002	12	1	-	-	-

Table 10: Collection Data of Praying mantid *H. bipapilla* on Mulberry for 2000-2002 at Kagal

Month	*Date of Collection*	*Adult Predators*	*Predator Nymphs*	*No. of Oothicae*	*No. of Oothicae Parasitized*	*Unknown Mortality (%)*
September	03-09-2000	34	23	4	1	1
	17-09-2000	26	19	1	0	1
October	01-10-2000	24	15	-	-	-
	15-10-2000	12	12	2	1	-
	29-10-2000	17	8	-	-	-
November	12-11-2000	14	5	-	-	-
	26-11-2000	9	2	-	-	-
December	10-12-2000	3	-	-	-	
	24-12-2000	4	-	-	-	
January	07-01-2001	6	-	-	-	
	21-01-2001	4	-	-	-	
February	04-02-2001	6	1	-	-	-
	18-02-2001	7	-	-	-	
March	04-03-2001	8	-	-	-	-
	18-03-2001	6	-	-	-	
April	08-04-2001	9	2	-	-	-
	22-04-2001	4	-	-	-	
May	06-05-2001	1	-	-	-	
	20-05-2001	2	-	-	-	
June	03-06-2001	11	13	-	-	-
	17-06-2001	12	5	-	-	-
July	01-07-2001	14	22	2	0	1
	15-07-2001	16	8	-	-	-
	29-07-2001	17	6	-	-	-
August	12-08-2001	19	8	1	-	-
	28-08-2001	20	10	-	-	-
September	09-09-2001	31	40	4	1	2
	23-09-2001	25	18	3	-	-

Contd...

Table 10–Contd...

Month	Date of Collection	Adult Predators	Predator Nymphs	No. of Oothicae	No. of Oothicae Parasitized	Unknown Mortality (%)
October	07-10-2001	23	6	-	-	-
	21-10-2001	9	-	-	-	
November	04-11-2001	17	4	1	-	-
	18-11-2001	8	-	-	-	
December	02-12-2001	3	-	-	-	
	16-12-2001	6	-	-	-	
	30-12-2001	4	-	-	-	
January	13-01-2002	2	-	-	-	
	27-01-2002	5	-	-	-	
February	10-02-2002	6	-	-	-	
	24-02-2002	8	-	-	-	
March	16-03-2002	10	-	-	-	
	24-03-2002	12	3	-	-	-
April	07-04-2002	9	-	-	-	
	21-04-2002	6	-	-	-	
May	05-05-2002	2	-	-	-	
	19-05-2002	4	-	-	-	
June	02-06-2002	10	2	-	-	-
	16-06-2002	11	-	-	-	
	30-06-2002	12	-	-	-	
July	14-07-2002	14				
	28-07-2002	17	20	2	0	-
August	11-08-2002	16	13	-	-	-
	25-08-2002	19	7	1	-	-
September	09-09-2002	24	39	3	1	-
	22-09-2002	31	18	2	-	-
October	06-10-2002	22	8	-	-	-
	20-10-2002	19	10	1	-	-
November	03-11-2002	17	2	-	-	-
	17-11-2002	11	-	-	-	-

Table 11: Collection Data of Praying mantid *H. bipapilla* on Mulberry for 2000-2002 at Kolhapur

Month	*Date of Collection*	*Adult Predators*	*Predator Nymphs*	*No. of Oothicae*	*No. of Oothicae Parasitized*	*Unknown Mortality (%)*
September	03-09-2000	32	18	4	1	1
	17-09-2000	26	28	1	-	-
October	01-10-2000	24	4	-	-	-
	15-10-2000	22	15	1	-	-
	29-10-2000	17	3	-	-	-
November	12-11-2000	15	2	-	-	-
	26-11-2000	11	5	-	-	-
December	10-12-2000	6	-	-	-	
	24-12-2000	4	1	-	-	-
January	07-01-2001	5	-	-	-	
	21-01-2001	8	-	-	-	
February	04-02-2001	7	-	-	-	
	18-02-2001	8	-	-	-	
March	04-03-2001	11	2	-	-	-
	18-03-2001	10	-	-	-	
April	08-04-2001	13	1	-	-	-
	22-04-2001	5	-	-	-	
May	06-05-2001	2	-	-	-	
	20-05-2001	3	-	-	-	
June	03-06-2001	12	5	-	-	-
	17-06-2001	13	16	2	-	-
July	01-07-2001	15	9	-	-	-
	15-07-2001	17	14	1	-	-
	29-07-2001	18	12	-	-	-
August	12-08-2001	20	16	-	-	-
	28-08-2001	20	7	-	-	-
September	09-09-2001	33	22	4	2	1
	23-09-2001	26	14	2	-	-

Contd...

Table 11–Contd...

Month	*Date of Collection*	*Adult Predators*	*Predator Nymphs*	*No. of Oothicae*	*No. of Oothicae Parasitized*	*Unknown Mortality (%)*
October	07-10-2001	24	10	-	-	-
	21-10-2001	20	8	1	-	-
November	04-11-2001	18	3	-	-	-
	18-11-2001	9	-	-		
December	02-12-2001	3	-	-	-	
	16-12-2001	4	-	-	-	
	30-12-2001	2	-	-	-	
January	13-01-2002	5	-	-	-	
	27-01-2002	6	-	-	-	
February	10-02-2002	7	-	-	-	
	24-02-2002	9	-	-	-	
March	16-03-2002	10	-	-	-	
	24-03-2002	12	1	-	-	-
April	07-04-2002	11	-	-	-	
	21-04-2002	6	-	-	-	
May	05-05-2002	3	-	-	-	
	19-05-2002	4	-	-	-	
June	02-06-2002	12	-	-	-	
	16-06-2002	12	-	-	-	
	30-06-2002	14	-	-	-	
July	14-07-2002	16	-	-	-	
	28-07-2002	19	20	-	-	-
August	11-08-2002	18	-	-	-	
	25-08-2002	20	32	3	-	-
September	09-09-2002	25	55	3	1	1
	22-09-2002	32	14	1	-	-
October	06-10-2002	23	6	-	-	-
	20-10-2002	20	8	1	-	-
November	03-11-2002	18	2	-	-	-
	17-11-2002	13	1	-	-	-

Chapter 4
Habitat Identification

Habitat of an organism is the place where it lives (Hodek, 1953). Many biocontrol agents of pest live in very close association with their host/preys. Shelter provision has been devised for encouragement and conservation of predators of pests (Hartman, 1971). The loss of shelter, nesting and roosting sites of predators seriously affect the population dynamics of predatory species. Hence Holdren & Ehrlich (1974) says that "productivity is achieved at the expense of stability". Shrubs, hollow trees, brush piles, orchards, forests etc. are good sources of predator shelter for both vertebrate and invertebrate groups. The residual manure harbours a sufficient number of natural enemies (predators/parasitoids). In Germany and many European countries the fabricated shelters and nest boxes are provided to predators for their encouragement and suppression of pest populations (Franz, 1961; Hagen & Franz, 1973; Buckner, 1966). Several insect predator species like *Polistes* spp. (paper wasps) which predate on Tobacco horn worm *Manduca* spp. are encouraged by supplementing them nesting shelters. However, it is well known fact that the modern intensive agro-ecosystems tend to eliminate suitable nesting sites (Lawson *et al.*, 1961).

In general, environmental diversity favours the conservation and increase the number of predators (Glen, 1954). Floral diversity is most easily realized in the forest environment (Graham, 1928). According to Bess *et al.* (1947) healthy and diverse forest structure encourages the predators specially small mammals and insect predators. However, in other agro-ecosystems, diversity may be more difficult to obtain because of efficiency benefits derived from the practices of monoculture (PSAC, 1965).

Clean cultivation, destruction of crop residues and elimination of weedy and waste areas that provide alternate host plants, etc. are not always advantageous. If the residues are plowed under, or the debris and weeds destroyed before their emergence in spring, then many biocontrol agents will be lost. The so called "dirty field technique" may be advantageous for encouragement and conservation of predators (Robb *et al.*, 1974).

The shifting of agricultural practices toward a more complex polyculture and choice of plant mixtures, rotation schedules and other cultural practices should be aimed to increase diversity for intelligent use for encouraging predators (Southwood & Way, 1970; Sathe & Bhosale, 2001).

The predators in general are more abundant where there is good source for their preys (Sathe & Bhosale, 2001). On an annual crops, orchards, in persistent agro-ecosystem etc, the predators are abundant. Generally, predators like the habitat of their preys. The quality of habitat has relation with the availability of predators. Predator habitat identification has tremendous importance in biological pest control. However, no attention has been paid on this important aspect in India. Keeping in view the above facts the present topic was selected.

An extensive survey was made for identification of habitats of predators in Kolhapur district during the years 2000-2002. The Tahasils Hatkanangale, Kagal, Radhanagari of Kolhapur district (Fig. 12) were selected for the identification of habitat of predators. Observations on the characteristics of habitat were made by visiting above places 15 days interval throughout the years, 2000-2002. Related flora and fauna in the habitat have been identified by consulting appropriate literature/experts. Other specialized features of the habitat were also located by spot observations.

Various types of habitats of predators located in Kolhapur districts of four centres namely, Kagal, Hatkanangale, Kolhapur, and Radhanagari are shown in the Figures 34 to 72. The habitats found in different places of Kolhapur district and their characteristics are given below. The following agro-ecosystems have been visualized as a habitat of predators in four centres.

(I) Kagal Center

1. Jowar (Fig. 40)
2. Mulberry
3. Teak (Fig. 35)
4. Subabul
5. Bamboo (Fig. 37)
6. Gallardia ecosystem (Fig. 36)
7. Sugarcane + Sunhemp (Fig. 41)
8. Other agro-ecosystems of various crops like Groundnut, Maize, Soybeans (Fig. 38 & 44), Tobacco (Fig. 43), Sugarcane (Fig. 39), Cotton (Fig. 42), Cabbage (Fig. 45), Rice and mixed agro-ecosystem (Fig. 34) etc.

(II) Hatkanangale Center

1. Jowar (Fig. 46)
2. Sugarcane
3. Teak
4. Bamboo (Fig. 49)
5. Subabul
6. Other agro-ecosystems like Maize, Groundnut, Tobacco, Tomato (Fig. 50), Brinjal (Fig. 48), Beans, Cabbage (Fig. 47), and Chilli (Fig. 51).

(III) Kolhapur Center

1. Jowar
2. Mulberry
3. Bamboo (Fig. 54)
4. Teak (Fig. 62)
5. Subabul (Fig. 53 & 58)
6. Calotropis (Fig. 55)
7. Gallardia eco-system (Fig. 52)
8. Other agro-ecosystems like Groundnut (Fig. 63), Sugarcane (Fig. 59), Rice, Cabbage (Fig. 57), Guava (Fig. 56), Hycacinth Bean (Fig. 60), China Rose (Fig. 61), Kidney bean (Fig. 65), Custard apple (Fig. 66) and grassland eco-system (Fig. 64) etc.

(IV) Radhanagari Center

1. Sugarcane (Fig. 67)
2. Cabbage
3. Teak (Fig. 72)
4. Bamboo (Fig. 69)
5. Other agro-ecosystems Rice (Fig. 68), Kidney bean (Fig. 71) and Jowar (Fig. 70).

The faunal observations made on the habitats of predators in different centers of study are recorded in Table 12.

In mixed agroecosystem, diversity of preys found increased in various centres in Kolhapur district. Aphids (*Macrociphum pisi* Kalt.) on Hycacinth bean (Fig. 75) and Kidney bean (Figs. 77, 78 & 81) were predated by coccinellid beetles. In addition, it has been observed that calotropis aphids (*Aphis nerii* B.F.) were predated by *Menochilus* larvae (Fig. 82), Rose aphid (*Macrosiphum rosae* Linneus) (Fig. 102), *Gallaridia* aphid (*Aphis craccivora* Glover) (Fig. 76) and Mulberry white fly (*Aleurodicus dispersus* Russel) (Fig. 74) predated by *Menochilus* larvae. Psysillid on subabul (Fig. 73) predated by *Coccinella* larvae. Mulberry white fly (*A. dispersus*) also predated by *Scymnus* sp. of lady bird beetle. *Coccinella* and *Illeis* species predated on tobacco aphids (*Myzus persicae* Sulzer) (Fig. 80) and Gawar aphid (Fig. 79). Grapes mealy bugs (*Pseudcoccus* sp.) (Fig. 99) predated by *Cryptolaemus* sp., custardapple mealy bugs (*Pseudococcus* sp.) (Fig. 101) predated by praying mantid (*Creobroter marathi*). China Rose scale (*Puiveneria* sp.) (Figs. 87 & 103) predated by *Hierodula* sp. Nymph of praying mantid feeding on tiger butterfly (Figs. 83 & 90) and praying mantid feeding behaviour are shown in Figs. 84, 88, 91 & 92.

Legume crops are important source of many predators (Sathe & Bhosale, 2001). *Chrysopa* sp., *Geocoris* sp. Lady bird beetles, syrphids etc. are found in wheat, jowar and barley agro-ecosystems. According to Sathe & Bhosale (2001) mulberry is good source for predators like praying mantids and coccinellid beetles. Nabids are associated with corn crop since a very large number of aphids found feeding upon the corn.

The aphids is good source of food for many insect predators. The aphid *A. craccivora* is found feeding on more than 18 families of

Plate VIII

Fig. 52: Gallardia Ecosystem (Kolhapur), Fig. 53: Subabul (*Leucaena retusa* B.) Ecosystem (Kolhapur), Fig. 54: Bamboo (*Dendrocalamus strictus*) Ecosystem (Kolhapur), Fig. 55: Calatropis *gignatea* (Habitat) (Kolhapur), Fig. 56: Guava *Psidium guajava* (Habitat) (Kolhapur).

Plate IX

Fig. 57: Cabbage *Brassica oleracea* agro-ecosystem (Kolhapur), Fig. 58: Subabual (*Leucaena retusa* L.) Ecosystem (Kolhapur), Fig. 59: Sugarcane *Saccharum officinarum* Agro-ecosystem (Kolhapur), Fig. 60: Hycacinth Bean *Lablab* (*Dolichos*) *purpureus* with Aphids and Coccinellids Larvae (Kolhapur), Fig. 61: China Rose (*Hibiscus rosasinensis* L.) (Habitat) (Kolhapur).

Plate X

Fig. 62: Teak (*Tectonia grandis*) Ecosystem (Kolhapur), Fig. 63: Groundnut (*Arachis hypogea*) Ecosystem (Kolhapur), Fig. 64: Grassland Ecosystem and Oothica (Kolhapur), Fig. 65: Kidney bean *(Phaseolus vuglare* L.) (Habitat) (Kolhapur), Fig. 66: Custard Apple (*Anona squamosa*) (Habitat) (Kolhapur).

Plate XI

Fig. 67: Sugarcane (*Saccharum officinarum*) agro-ecosystem (Radhanagri), Fig. 68: Rice (*Oryzus sativa*) agro-ecosystem (Radhanagri), Fig. 69: Bamboo (*Dendrocalamus strictus*) Ecosystem (Radhanagri), Fig. 70: Jowar (*Sorghum vulgare*) (Habitat) (Radhanagri), Fig. 71: Kidney bean (*Phaseolus vulgaris* L.) (Radhanagri), Fig. 72: Teak (*Tectonia grandis*) Ecosystem (Radhanagri).

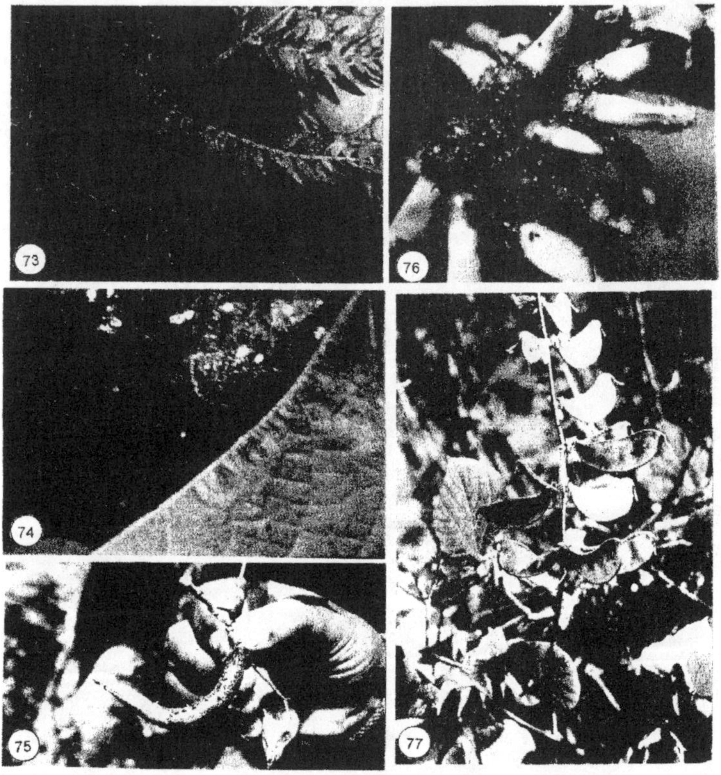

Plate XII

Fig. 73: Physillids on Subabul (*Leucaena retusa* B), Fig. 74: Whiteflies on Cotton Leaf (*Gossypium hirsutum*), Fig. 75: Aphids and Coccinellids as Hycacinth Bean [*Lablab* (*Dolichos*) *purpureus*], Fig. 76: Aphids and Coccinellid Grubs on *Gallaradia bulchella*, Fig. 77: Aphids and Coccinellid Larvae on Kidney bean (*Phaseolus vulgaris*).

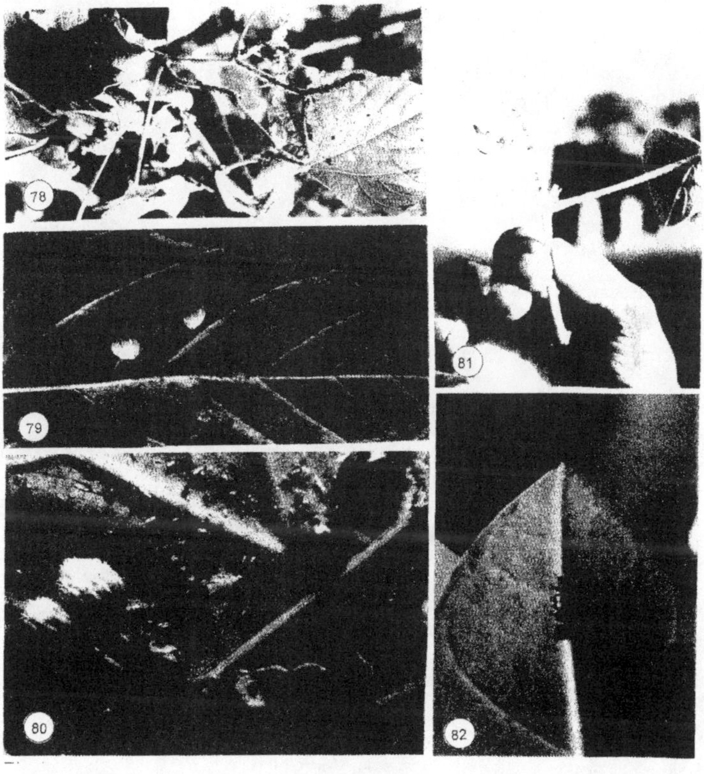

Plate XIII

Fig. 78: Coccinellids Larvae Feeding on Kidney bean (*Phaseolus vulgare*) Aphid, Fig. 79: *Illeis* with Grub, Pupae and Adults on Gawar (*Cymaposis tetragonoloba* L.) Leaf, Fig. 80: Pupae and Adult *Coccinella* Beetles on Tobacco Leaf (*Tabaccum tabici*), Fig. 81: Lady Bird Beetle Larvae and Aphid Population on Kidney bean (*Phaseolus vulgaris*), Fig. 82: Coccinellid Larva Feeding on *Calotropis* Aphid.

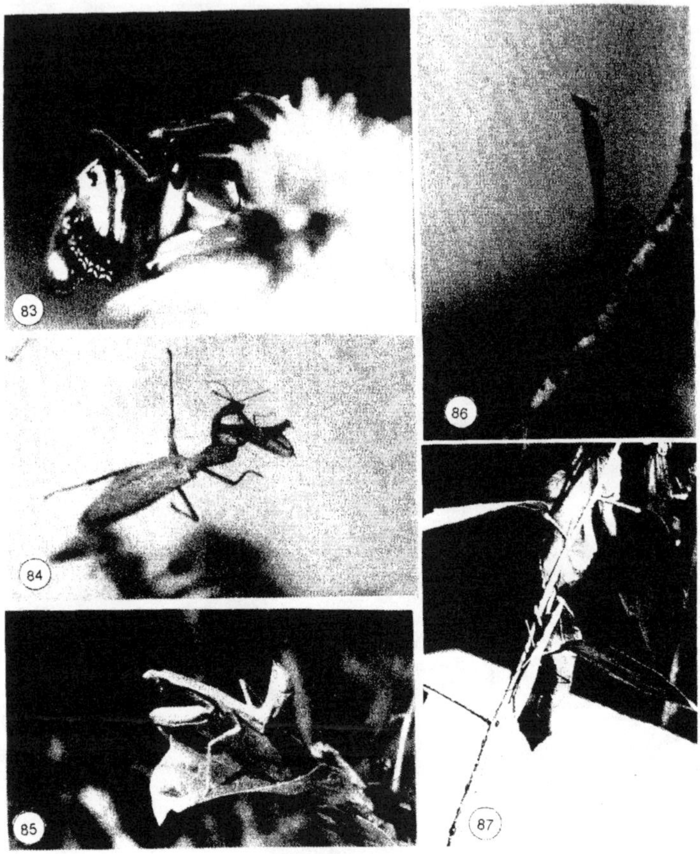

Plate XIV

Fig. 83: Mantid Nymph Feeding on Tiger Butterfly, Fig. 84: *Creoboter* Mantid Feeding on Grasshopper, Fig. 85: Nymph of Praying Mantid Searching for Homopterous Preys, Fig. 86: Nymph of *Schizocephala* Watching Homopterous Preys on Mulberry *(Morus alba)* System, Fig. 87: Praying Mantid *Hierodula* sp. Feeding on Scale on China Rose (*Hibicus rosasinensis* L.).

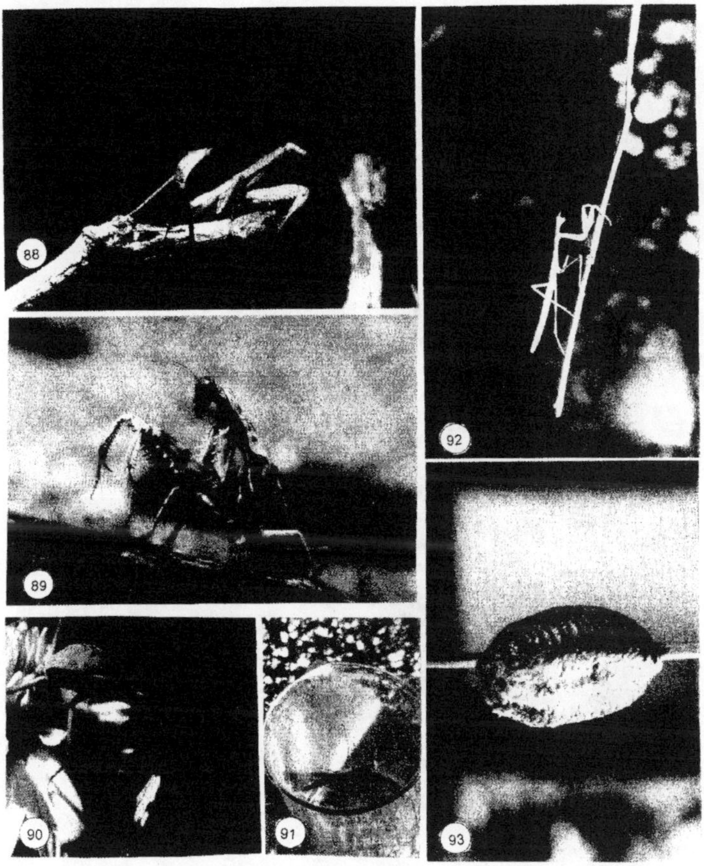

Plate XV

Fig. 88: Mantid Nymph Searching for Prey, Fig. 89: Mantid (*Elmantis nira*) in Position of Catching the Prey, Fig. 90: Nymph Feeding on Tiger Butterfly, Fig. 91: Mantid *Humberticella*, Fig. 92: Mantid Nymph (*Schizocephala* sp.) Watching Prey, Fig. 93: Oothica of Praying Mantid.

Plate XVI

Fig. 94: *Heliothis* on Cotton (*Gossypium hirsutum* L.), Fig. 95: Spotted Bolloworm on Cotton (*G. hirsutum*), Fig. 96: Cotton Semilooper on Jute (*Corchorus* sp.), Fig. 97: Wooly Aphid on Sugarcane (*Saccharum officinarum*), Fig. 98: Cockroach (*Periplanata americana*).

Plate XVII

Fig. 99: Mealy Bugs on Grapes (*Vitus venifera*), Fig. 100: Wooly Aphids on Sugarcane Leaf (*Saccharum officinarum*), Fig. 101: Mealy Bugs on Custard Apple (*Annona squamosa* L.), Fig. 102: Aphids on Rose (*Rosa* sp.), Fig. 103: Scales on China Rose (*Hibiccus rosasinensis* L.).

Table 12: Characteristics of Habitats of Predators

Sl. No.	Study Center	Predator	Vegetation (Agro-ecosystem)	Average Rainfall in mm	Average Temp. (°C)	Average Humidity (%)	Preys
(1)	(2)	(3)	(4)	(5)	(6)	(7)	(8)
1.	Hatkanangale	1. *Menochilus sexmaculatus* (Fab.)	Jowar (*Sorghum vulgare* Pers.)	87.5	32.5	77	Aphid, *Rhophalosiphum maidis*, (Fitch.) Delphacid *Perigrinus maidis* (Ashmed)
		2. Praying mantid (*Hierodula bipapilla*) (Serville - Audinet)	Jowar (*S. vulgare*)	-do-	-do-	-do-	*Helicoverpa armigera* (Hubn.) larvae, Delphacid *P. maidis*
		1. *M. sexmaculatus*	Sugarcane (*Saccharum officinarum* L.)	-do-	-do-	-do-	Sugarcane aphid, *Aphis sacchari* (Z.) Mealy bugs, *Saccharlcoccus saccharl* (Cock.)
		1. *Jauravia* (Clanis) *pallidula* (Mots)	Teak *Tectonia grandis* (L.)	-do-	-do-	-do-	Scale insect
		2. *Scymnus caccivora* (Diyar)	-do-	-do-	-do-	-do-	*Aphis* sp.
		1. *Anisolemnia dilatata* (Fab.)	Bamboo, *Bambusa bambu*	-do-	-do-	-do-	*Oregma* sp.
		2. *Carla dilatata* (Fab.)	-do-	-do-	-do-	-do-	*Oregama* sp.
		3. *J. pallidula*	-do-	-do-	-do-	-do-	Scale insect

Contd....

Table 12 – contd....

(1)	(2)	(3)	(4)	(5)	(6)	(7)	(8)
		4. *Nephus* sp.	Bamboo, *Bambusa bambu*	87.5	32.5	77	*S. sacchari*
		5. *Sticholotis criberellata* (Fair)	-do-	-do-	-do-	-do-	Hard scales
		1. *Chilocorus bijugus* (Mul.)	Mulberry (*M. alba*)	-do-	-do-	-do-	*Coccus* sp.
		2. *Cryptolaemus montrouzieri* (Mul.)	-do-	-do-	-do-	-do-	*Maconellicoccus hirsutus* (Green)
		3. *Pharoscymnus horni* (Weise)	-do-	-do-	-do-	-do-	*Diaspri echninocacti*
		4. Praying mantid (*H. bipapilla*)	-do-	-do-	-do-	-do-	*Spilosoma obliqua*,(Wlk) *Spodoptera litura* (Fab), jassids, aphids, etc.
		1. *M. sexmaculatus*	Cabbage, (*Brassica oleracea* var. *capitata*)	-do-	-do-	-do-	*R. maidis*
		2. *Illeis darbari* (S. & B.)	-do-	-do-	-do-	-do-	*Aphis gossypi* (Glover)
		3. *Coccinella transversalis* (Fab)	-do-	-do-	-do-	-do-	*Myzus persicae* (Sulz)
		4. *Harmonia octomaculata* (Fab)	-do-	-do-	-do-	-do-	*Brevicorne brassicae* (Linn.)
2. Kagal		1. *M. sexmaculatus*	Jowar, (*S. vulgare*)	100.0	32.4	72	Aphid, *R. maidis*
		2. *Coccinella septampunctata* L.	-do-	-do-	-do-	-do-	Aphid, *R. maidis*

Contd....

Table 12 – contd....

(1)	*(2)*	*(3)*	*(4)*	*(5)*	*(6)*	*(7)*	*(8)*
		3. *Harmonia* sp.	Jowar, (*S. vulgare*)	100.0	32.4	72	Aphid, *R. maidis*
		4. *I. darbari* (S&B)	-do-	-do-	-do-	-do-	Aphid, *R. maidis*
		5. *Menochilus shivajensis* (S&B)	-do-	-do-	-do-	-do-	Aphid, *R. maidis*
		6. *Scymnus marathi* (S&B)	-do-	-do-	-do-	-do-	*Dactynotus compositae* (Theobald)
		7. *Vernia polyphagi* (S&B)	-do-	-do-	-do-	-do-	*R. maidis*
		8. Tiger beetle, *Chlaenius rayotus*,	-do-	-do-	-do-	-do-	*H. armigera*
		Anthia sexguttata	-do-	-do-	-do-	-do-	Grasshopper, *Heiroglapus banian* (Fab.)
		1. *C. bigjugus*	Mulberry, (*M. alba*)	-do-	-do-	-do-	*Coccus* sp.
		2. *C. montrouzieri*	-do-	-do-	-do-	-do-	*M. hirsutus*
		3. *Pullus elegans*	-do-	-do-	-do-	-do-	*Pulvinaria* spp.
		4. *P. horni*	-do-	-do-	-do-	-do-	*Pseudococcus* sp.
		5. Praying mantid (*H. bipapilla*)	-do-	-do-	-do-	-do-	*S. litura,* leaf roller, jassids, scales etc.

Contd....

Table 12 – contd....

(1)	*(2)*	*(3)*	*(4)*	*(5)*	*(6)*	*(7)*	*(8)*
		1. *J. pallidula*	Teak, (*T. grandis*)	100.0	32.4	72	Scale insect
		2. *S. caccivora*	-do-	-do-	-do-	-do-	*Aphis* sp.
		1. *A. dilatata*	Bamboo (*B. bambu*)	-do-	-do-	-do-	*Oregma* sp.
		2. *C. dilatata*	-do-	-do-	-do-	-do-	*Oregma* sp.
		3. *J. pallidula*	-do-	-do-	-do-	-do-	Scale insect
		4. *Nedphus* sp.	-do-	-do-	-do-	-do-	*S. sacchari*
		5. *S. criberellata*	-do-	-do-	-do-	-do-	Hard scales
		1. *M. sexmaculatus*	Cabbage (*B. oleracea*)	-do-	-do-	-do-	*R. maidis*
		2. *I. darbari*	-do-	-do-	-do-	-do-	*A. gossypii*
		3. *C. transversalis*	-do-	-do-	-do-	-do-	*M. persicae*
		4. *H. octomaculata*	-do-	-do-	-do-	-do-	*P. brassicae*
3	Kolhapur	1. *M. sexmaculatus*	Jowar (*S. vulgare*)	112.5	31.5	73	Aphid, *R. maidis*
		2. *M. shivajensis*	-do-	-do-	-do-	-do-	*R. maidis*
		3. *C. septampunctata*	-do-	-do-	-do-	-do-	*R. maidis*
		4. *I. darbari*	-do-	-do-	-do-	-do-	*R. maidis*

Contd....

Table 12 – contd....

(1)	*(2)*	*(3)*	*(4)*	*(5)*	*(6)*	*(7)*	*(8)*
		5. *S. marathi*	Jowar (*S. vulgare*)	112.5	31.5	73	*R. maidis*
		6. *V. polyphagi*	-do-	-do-	-do-	-do-	*R. maidis*
		1. *C. bigjugus*	Mulberry (*M. alba*)	-do-	-do-	-do-	*Coccus* sp.
		2. *C. montrouzieri*	-do-	-do-	-do-	-do-	*M. hirsutus*
		3. *P. elegans*	-do-	-do-	-do-	-do-	*Pulvinaria* sp.
		4. *P. horni*	-do-	-do-	-do-	-do-	*Diaspri echninocacti*
		5. Praying mantid (*H. bipapilla*)	-do-	-do-	-do-	-do-	*S. obliqua, S. litura* leaf rollers, scales, jassids etc.
		1. *J. pallidula*	Teak (*T. grandis*)	-do-	-do-	-do-	Scale insect
		2. *S. caccivora*	-do-	-do-	-do-	-do-	*Aphis* sp.
		1. *A. diiatata*	Bamboo (*B. bambu*)	-do-	-do-	-do-	*Oregma* sp.
		2. *C. dilatata*	-do-	-do-	-do-	-do-	*Oregma* sp.
		3. *J. pallidula*	-do-	-do-	-do-	-do-	Scale insect
		4. *Nephus* sp.	-do-	-do-	-do-	-do-	*S. sacchari*
		5. *S. criberellata*	-do-	-do-	-do-	-do-	Hard scales

Contd....

Table 12 – contd....

(1)	*(2)*	*(3)*	*(4)*	*(5)*	*(6)*	*(7)*	*(8)*
		1. *M. sexmaculatus*	Cabbage (*B. oleracea*)	112.5	31.5	73	*R. maidis*
		2. *I. darbari*	-do-	-do-	-do-	-do-	*A. gossypii*
		3. *C. transversalis*	-do-	-do-	-do-	-do-	*M. persicae*
		4. *H. octomaculata*	-do-	-do-	-do-	-do-	*P. brassicae*
4.	Radhanagari	1. *M. sexmaculatus*	Sugarcane (*Saccharum officinarum*	250.0	31.0	71	Sugarcane aphid *A. sacchari*
							Mealy bugs *S. sacchari*
		1. *J. pallidula*	Teak (*T. grandis*)	-do-	-do-	-do-	Scale insect
		2. *S. caccivora*	-do-	-do-	-do-	-do-	*Aphis* sp.
		1. *A. dilatata*	Bamboo (*B. bambu*)	-do-	-do-	-do-	*Oregma* sp.
		2. *C. dilatata*	-do-	-do-	-do-	-do-	*Oregma* sp.
		3. *J. pallidula*	-do-	-do-	-do-	-do-	Scale insect
		4. *Nephus* sp.	-do-	-do-	-do-	-do-	*S. sacchari*
		5. *S. criberellata*	-do-	-do-	-do-	-do-	Hard scales
		1. *M. sexmaculatus*	Cabbage (*B. oleracea*)	-do-	-do-	-do-	*R. maidis*
		2. *I. darbari*	-do-	-do-	-do-	-do-	*A. gossypii*
		3. *C. transversalis*	-do-	-do-	-do-	-do-	*M. persicae*
		4. *H. octomaculata*	-do-	-do-	-do-	-do-	*P. brassicae*

plant kingdom. However, they are more abundant on many plants of the family polygoneaceae. Thus, these plants are good source for predators specially coccinellid beetles. *Menochilus sexmaculatus, Coccinella septumpunctata, Brumus* sp. are associated with above plants. Tobacco ecosystem in European countries is good source for a predator *Polister* spp. (Wasp) which predate upon tobacco horn worm *Manduca sextra* and other *Manduca* spp.

President's Science Advisory Committee (1965) have reported that floral diversity was important aspect for attracting predators. Sathe & Bhosale (2001) reported that the family compositae was potential source for many insect predators. They further recorded that vegetable crops, groundnut, safflower and many pulse crops were important sources for predators. The above crops were attacked by aphids, jassids, mealy bugs, scales, caterpillars etc. and further predated by the predators such as *Crysopa*, praying mantids, many coccinellids and plant bugs. Hence, these predators were associated with above crops. The cotton was very good source for predators like *Menochilus sexmaculatus, Coccinella* sp. *Scymnus* sp., *Chrysopa* sp. and several species of spider (Sathe & Bhosale, 2001).

In the present study, three coccinellids namely *M. sexmaculatus, C. transversalis* and *Harmonia octomaculata* (Fab.) were recorded in vegetable ecosystem in Kolhapur district. While in teak ecosystem, four species of coccinellids, *M. sexmaculatus, C. montrouzieri, S. (p.) coccivora* and *J. soror* have been reported. *I. purchasi, A. gossypi* and *E. vrgata* are hemipterous pests of teak plant. Hence above predators harboured on the teak.

Omkar and Bind (1993) reported some agricultural and horticultural plants as a source for six coccinellid beetles namely, *Coccinella septumpunctata* Linn., *Coccinella transversalis* Fab., *Menochilus sexmaculata* (Fab.), *Brumoides suturalis* (Fab.), *Micraspis* (*Verania*) *discolor* (Fab.) and *Micraspis cardoni* (Weise). Patro & Behera (1992) reported Bean (*Dolichos lablab*) as an important source for *Scymnus castaneus* Mulsant and *Scymnus pyrochellus* Mulsant from Orissa. While, Rawat *et al.* (1992) recorded apple as an important source for a lady bird beetle, *Chilocorus bijugus* Mulsant which predated San Jose scales, *Quadraspidiotus perniciosus* (Comstock) the potential pest of apple in hilly tracts of India.

From Malaysia van Vreden & Ahmadzabidi (1986) studied the habitat of predators. They reported that paddy agro-ecosystem was very rich source for several vertebrate and invertebrate predators.

The insect predator species observed in paddy ecosystem are listed below:

Damsel flies—Ischnsura senegalensis (Rambur), *Agriocnemis pygmaea* (Rambur),

Dragon flies—Crocothemis servilia (Drury), *Orthetrum sabina* (Drury), *Diplacodes trivialis* (Drury) and *Neurothemis tullia tullia* (Drury),

Lady bird beetles—Hamronia octomaculata (Fab.), *Micraspis discolor* (Fab.), *Micraspis inops* (Mulsant) and *Lemnia biplagiata* (Swartz),

Rove-beetle—Paederus fuscipes Curtis,

Ground beetle—Casnoidea indicci (Thunberg), *Casnoidea ishii* Habu, *Anoplogenius microgonus* (Chandoir), *Colliuris fuscipennis* (Chandoir), *Chlaenius bioculatus* (Chandoir),

Long horned grasshoppers—Concephalus longipennis (de Haan),

Sword bearing Crickets—Metioche vittaticollis (Stal.), *Anaxipha* sp.,

Ripple bugs—Microvelia douglasi Scott,

Shield bugs—Andrallus spinidens (Fab.),

Plant bugs—Cytorhinus lividipennis Reuter.,

Water striders—Limnogonus fossarum (Fab.),

Assassin bugs—Scipinia rapax Miller, *Euagoras plagiatus* Burmeister, *Polytoxus selanorensis* Miller, *Sirthenea flavipes* (Stal.), *Ectrychotes crudelius* Fab., *Coranus fuscipennis* Reuter, *Oncocephalus impudicus* Reuter, *Peregrinator biannulipes* (M.& S.),

Ants—Monomorium destructor Jerdon, *Camponotus* sp., *Solenopsis germinata* Fab.,

Wasps—Ropalidia fasciata Fab., *Ropalidia marginata sundaica* v.d. Vecht., *Sceliphron madraspatanum madraspatanum* (Fab.),

Chloropid flies—Anastrichus pygmaeus (Lamb), *Poecilotraphera taeniata* (Macquart)

Earwigs—Proreus simulans (Stal.).

In the present study, the important component of habitat was vegetation. The predators such as lady bird beetles and praying mantids were largely recorded on agricultural and horticultural plants. The teak forest ecosystem was mostly attractive for lady bird beetles. Likely, bamboo ecosystem was also good source for the predaceous coccinellids.

Many times small populations of coccinellids, *M. sexmaculatus* were observed on congress grass, *Parthenium hysteroforus* Linn.

An ideal habitat should provide maximum facilities as basic needs of an organism, *i.e.* food, shelter and mate. However, the quality of habitat affect the population, survival and occurrence of the predators.

The availability of a reliable supply of food for each feeding stage in the predator's life cycle, at the proper time of the year is essential feature of the environment necessary for the encouragement of predators. Predation by syrphid larvae on cabbage aphids, *Brevicoryne brassicae* (L.), infesting Brussel's sprouts was greatest near the edges of the fields where wild flowers were most numerous (Pollard, 1968). Some predators like mirid bugs, feed on insect pests in the crop, but supplement their diet with plant food from weeds in nearby hedgrews, particularly in the absence of sufficient prey. Van Emden (1965a) called this situation as "bridge for failures in the synohronization of biocontrol agent". Carlson & Chiang (1973) reported that the "food spray" stimulates honeydew and/or pollen, and is utilized by lacewings, *C. carnea* Stephens, and several species of syrphidae and coccinellidae. A predatory ant, *Oecophylla longinoda* (Latreille), is heavily dependant on honeydew from a scale insect, *Saissetia zanzibarensis* Williams (Way, 1954). Hambleton (1944) reported that a change in the cropping system increased the diversity of a crop plants grown and, in turn, provided many alternate host insects for the build-up and maintenance of a large complex of polyphagus natural enemies which attacked. *Heliothis virescens* (F.).

Graham (1928) found that better drained portions of tamarack swamps supported a more diverse plant community than wet sites. In such areas, small mammal populations were higher and the existing larch saw fly *P. erichsonii* (Hartig) cocoon population was subject to much heavier predation. Van Emden (1965) reported that the measures used to discourage high pest densities may be more detrimental to predators, and in the long run may be ill-advised. Rabb *et al.* (1974) reported that many parasitoids and invertebrate predators pass the winter in debris, or within their host in crop residues or weedy alternate host plants. If the residues are plowed under, or debris and weeds destroyed before their emergence in spring, then many predatory organisms get lost. In the present study as a component of habitat emphasis is given on vegetation.

The dependance of any species on a given habitat was mainly determined by the occurrence of essential prey and the physical conditions of the environment (Hodek, 1973). For a number of species, the microclimatic conditions may be of primary importance. Certain strata of vegetation are preferred by certain coccinellid species IPerti (1965) reported that in south eastern France *C. septumpunctata* and *Semiadalia undecimnotata* preferred plants lower than 0.5 m. while, *Adalia bipunctata* and *Synhannonia conglobata* prefer trees above 2 m. Haunvik & Frye (1969) reported that *Hippodamia tredecimpunctata* and *H. convergens* were the most abundant species of all adult coccinellids on sweet corn in North Dakota. Putman (1957) recorded ten species of "larger aphidophagus coccinellids" in the Peach orchards of the Niagara, Canada. Dyadechko (1954) reported that the cereals was the import source for many species of coccinellids in Ukraine.

The suitability of a habitat is also dependent on seasons other than the presence of food or an adequate climate. Some plants may exert a mechanically deleterious effect on the predators. Plaut (1965) observed that the coccinellid reproduced in several habitats such as apple, cotton, orchards, watermelon and sugar-beet fields, but on bush beans only adults without any larvae could be found. He concluded that the hooked trichomes on bush beans were deleterious for the larvae.

Carabid habitats are divided into two groups: Wet, associated with surface water and dry. The former may be either Type - A: well developed vegetation and litter of humus layers, represented by bags, marshes and some - wood land, or type - B: Mud, Shingle or boulders at the margins of water bodies.

The dry habitats were classed as woodland, arable or grassland and the last including all other open habitats - such as pasture and moor. The coccinellids and praying mantids are associated with woodland and grassland habitats and tiger beetle with grassland and open habitat.

Greenslade (1968) reported that the complex of low temperatures and high humidities may be limiting for carabidae. Greenslade (1964) also reported that the desiccation risk in open habitats was shown to limit the distribution of species. The present observations on carabid beetles are in agreement with Greenslade (1964). Darlington (1943) stated that within dry habitats there was elimination of diurnal carabidae and concentration of nocturnal species, among which darker colour and - wings are common. The tiger beetle, studied in the present work, was diurnal and also nocturnal. Many times these beetle found rushing in the human dwelling at night. The tiger beetles were found in grassland and open habitat. Its prey consists small insects like crickets, silverfishes, leaf hoppers, small caterpillars, grasshoppers, etc.

Pearson (1964) studied the analogy of *Gerris* sp. (Hemiptera) and reported that the - winged form of *Bembidion grapei* Gyllenhal might be a cold - resistant low temperature phenomenon. The tiger beetle studied in the present work is winged and was abundant through out the year. However, its incidence was declined little but in cold season. Darlington (1963) put forward a very similar suggestion, related to the prevalence of winged carabid beetles. Brinkhurst (1959, 1963) stressed the importance of wing condition in relation to migratory potential. Greenslade (1968) studied the habitat and altitude distribution of carabidae (Coleoptera) in Argyll, Scotland and reported that at higher altitudes the mountain climate is responsible for further selective elimination of diurnal species, but the complex of low temperatures and high humidities may become limiting for most carabids. He further reported that diurnalism and fully developed flying wings were correlated. He also noticed that carabid fauna in a district in the S.W. Scottish Highlands, the number of species declined steadily with altitude 2500 feet, where there was an abrupt qualitative change to small montane fauna. Kolhapur's altitude is 601 to 650 m and most of spots selected in the present study comes under plain region. However the prevance of tiger beetle in Western Ghats would worthwhile study for further research.

A large number of praying mantis are associated with bushes, in grass, on various parts of trees such as flowers, leaves, stem, bark, etc. Mantis are most abundant upon bushes, and rare upon small crops. However, on mulberry garden, mantis population was sizable in Kolhapur district specially at Kolhapur, Kagal and Hatkanangale centre of the study. The number of mantids found limited in the sweep grass ecosystem. In Kolhapur district *H. bipapilla* was abundant in all the selected spots.

The mantid (*Schizocephala gramminae*) that live in grass were slender bodied and grass coloured like the grass, the dry tips of withered grass. Some species like *Creoboter* spp. and *Hierodula* spp. were leaf green and associated with leaves of mulberry, custard apple, and many other small bushes. Bark coloured species were associated with the bark of mulberry (Fig. 86), custardapple, guava, etc. They feed on whitefly larvae (Fig. 74), jassids, mealy bugs (Figs. 99 & 101), scales (Fig. 103), psyllids (Fig. 73), etc. Giglio Tos, (1927) reported that the Orchid mantis, *Gongylus gongloides*, which was floral simulator, the body, and wings so formed as to suggest a flower when a particular altitude is assumed. In these situations, the lower surface suggests, a blue flower, and insects coming to it were destroyed by the fore legs for feeding. Williams (1904) reported that the upper surface can be so arranged as to simulate an orchid flower, thus being primarily as a means of defense, the blue flower resemblance alone being used to obtain food. In general, the cryptic form and colour serves the double object of protecting the insect from toes and allowing it to be invisible to other insects which captured when they come within reach.

It is concluded that insect pest predators are not studied with respect to their habitats in India and there is endless scope to work on this aspect.

Chapter 5
Taxonomy

Taxonomy plays an important role in the fields of basic and applied biology including public health, environmental problems, national defense, wildlife management and Pest management in identifying the species and is an attractive profession nowadays. Taxonomists are needed in Universities, Research Institutions, Museums, Central and State Government agencies, Industries, Zoos etc. for identification of the species, to study the evolutionary history and guide the explorations for native and exotic species. The knowledge of an organisms relationship with the other permit us to develop means of control for target species more easily. Hence, pest control programme is totally dependent on correct identification of the target species (pests). Recently, increasing efforts have been directed towards the biological control programme of pests, since biological control is very effective alternative for pesticidal pest control. Biological pest control and taxonomy are interrelated and interdependent branches of science. There are several examples of failure of pest control programme due to incorrect identity of pests and pest biocontrol agents. Hence, in the present chapter Taxonomical studies have been made on Coccinellids, Carabids and Praying mantids, as these insects are very good biocontrol agents of insect pests of agriculture, forest and household.

Taxonomy of Coccinellid Beetles

The coccinellid beetles are familiar to people because of their prettiness and scientific importance in pest management. According to Lefroy (1909) out of 3,20,000 Coleopteran species in the world, 4,780 have been described from India. Amongst the various families of Coleoptera, Coccinellidae has outstanding importance since the family contain exclusively predaceous species (except epilachnids). They attack several insect pests. Hence, widely used in biological pest control (Omkar and Mishra, 2002). Coccinellids feed their larval and adult stages on aphids, scale insects, Psyllids, white flies, mealy bugs, phytophagous mites, etc. The above insects are injurious to various agricultural and forest crops. The Coccinellids have also great importance from the view point of evolutionary biology, population ecology, genetics, cytology, biogeography and biotechnology etc.

According to Sasaji (1971), about 490 genera and 4200 species of Coccinellidae have been reported from the world. Linnaeus (1758) for the first time described 36 species under the genus *Coccinella*. Later, Redtenbacher (1843) contributed on the sub division of the family Coccinellidae into aphidophagous and phytophagous groups. A monograph of Mulsant, "Securipalpus" is totally devoted to Coccinellids. Mulsant in 1850 established his system for identification for all the Coccinellids of the world. In his system he divided the family Coccinellidae into two groups "Gymnosomides" and "Trichosomides". His division was based on presence or absence of dorsal pubescene. Later, Croteh (1874), thoroughly revised the Mulsant's classification and established his own system. He combined platynaspis with Chilocorous and its allies and he treated apilachnids and Coccinellids to a single sub family Coccinellidae. However, Chapius (1876) revised the family Coccinellidae into two sub divisions, "Aphidophagous" and "Phytophagous" and aphidophagous sub division into 13 groups. During 1877 to 1925, Weise published more than one hundred papers on taxonomy of this family. He revised the family and gave a perspective of the systematic position of the genera.

During the years 2000-2002 very large number of specimen were collected from agricultural crops viz., Jowar, *Sorghum vulgare*, Pers.; Maize, *Zea mays* Linn.; Groundnut, *Arachis hypogaea* Linn.; Soybean, *Glycine soja* Linn.; Cow Pea, *Vigna catijang* Walp.; Sunflower, *Helianthus annus* Linn., Sunflower, *Carthumus tinctorius* Linn.; Cotton, *Gossypium hirsutum* Linn.; Mulberry, *Morus alba* Linn.; and vegetables, also from subabul, *Leucaena retusa*, Brew; Rui, *Calotropis gignatea* Linn.; China rose, *Hibiscus rosasinensis* Linn.; Composite garden plant, *Gaillardia bulchella* Fonger and Bamboo, *Dendrocalamus strictus* Roxb.; Teak *Tecktonia grandis*. The Coccinellid species described in this thesis

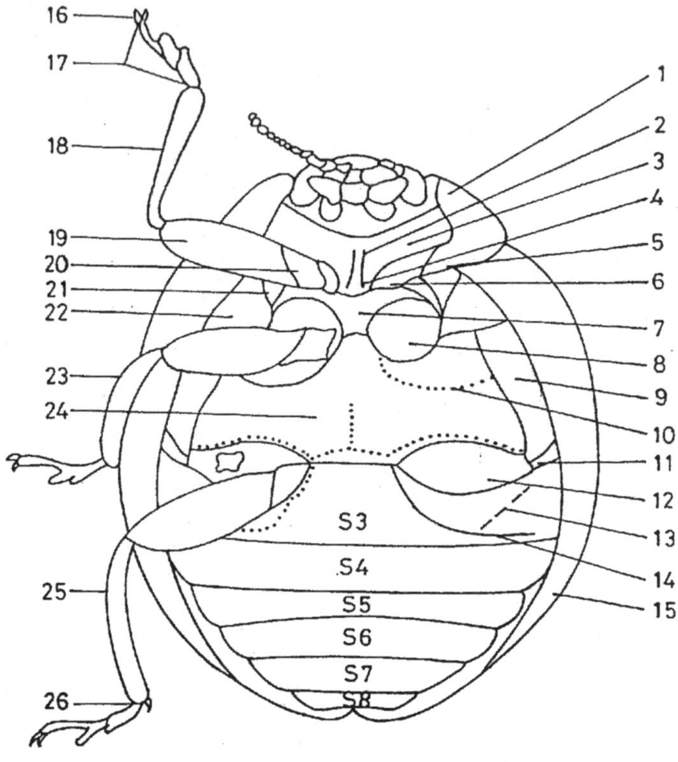

Plate XVIII

Fig. 104: Adult ventral view

(1) Hypomeron, (2) Prosternal carina, (3) Prosternum, (4) Prosternal process, (5) Postcoxal process, (6) Fore coxal cavity, (7) Mesosternum, (8 Middle coxal cavity, (9) Metepisternum, (10) Femoral line, (11) Metepimeron, (12) Hind coxal cavity, (13) Lateral femoral line, (14) Femoral line, (15) Elytral epipleuron, (16) Claw, (17) Tarsus, (18) Tibia, (19) Femur, (20) Trochanter, (21) Mesepisternum, (22) Mesepimeron, (23) Tibia of middle leg, (24) Metasternum, (25) Tibia of hind leg, (26) Tibiual spur of hind leg.

S8 : Sternite of abdominal segments

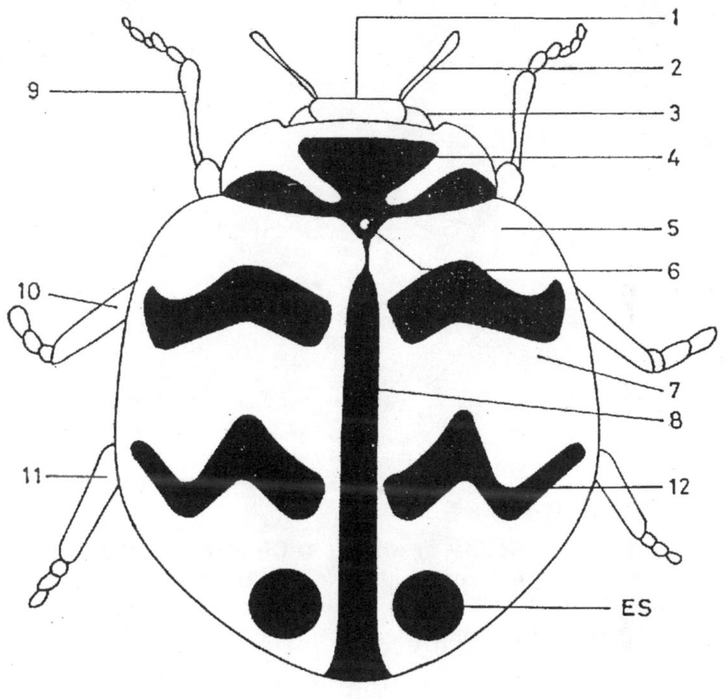

Plate XVIII

Fig. 105: Adult dorsal view

(1) Head, (2) Antenna, (3) Eye, (4) Pronotum, (5) Humeral callus, (6) Scutellum, (7) Elytra, (8) Elytral suture, (9) Fore leg, (10) Middle leg, (11) Hind leg, (12) Zig-Zag Marking.

ES: Elytral spots.

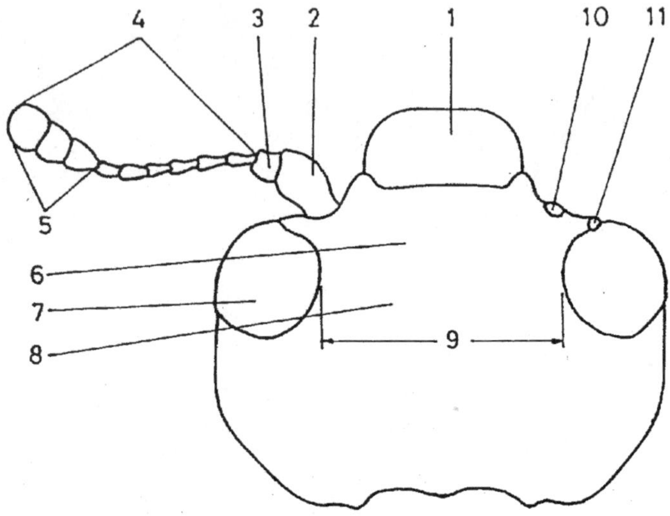

Plate XIX

Fig 106: Head dorsal view

(1) Labrum, (2) Scape of antenna, (3) Pedicel of antenna, (4) Flagellium of antenna, (5) Antennal club, (6) Clypeus, (7) Compound eye, (8) Frons, (9) Interocular distance, (10) Antennal socket, (11) Post antennal process.

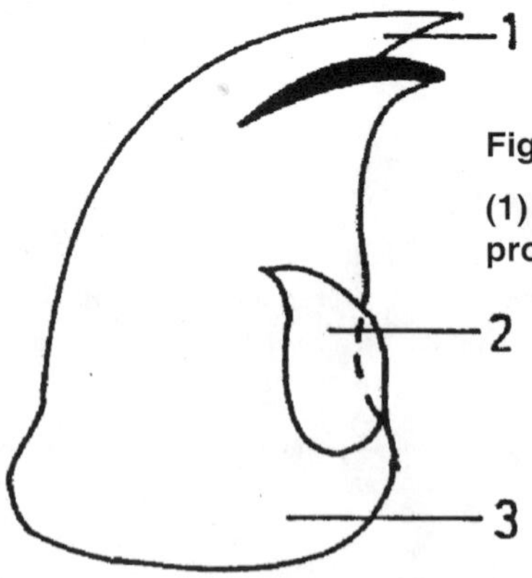

Plate XIX

Fig. 107: Mandible

(1) Apical tooth, (2) Membranous prostheca, (3) Basal tooth.

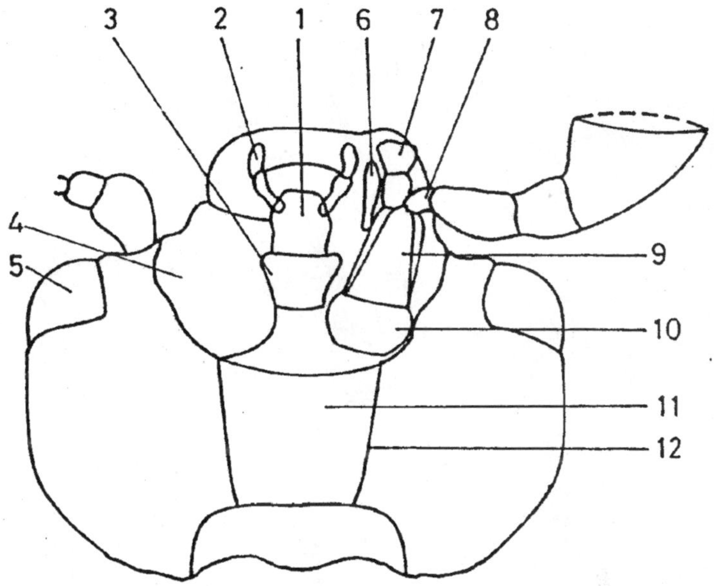

Plate XIX

Fig. 108: Head Ventral View.

(1) Prementum of labium, (2) Labial palp, (3) Mentum, (4) Tentorium, (5) Compound eye, (6) Lacinia of maxilla, (7) Galea of maxilla, (8) Maxillary palp, (9) Stipe of maxilla, (10) Cardo maxilla, (11) Gula, (12) Gular suture.

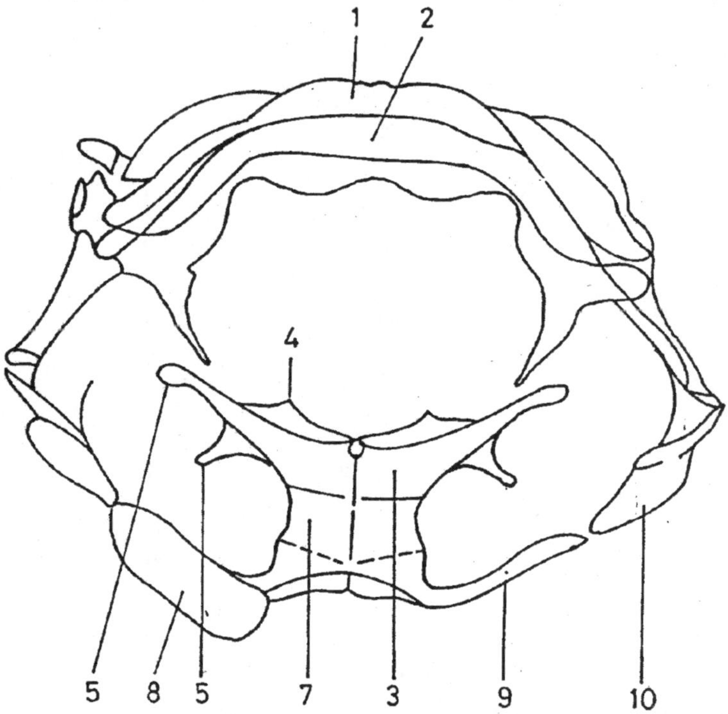

Plate XX

Fig. 109: Metathorax

(1) Metathorax, (2) Postnotum of metathorax, (3) Metendosternite, (4) Anterior tendon of metendosternite, (5) Lateral arm of metendosternite, (6) Lateral lobe of metendosternite, (7) Sternum, (8) Coxa of hind leg, (9) Metasternum, (10) Metepimeron.

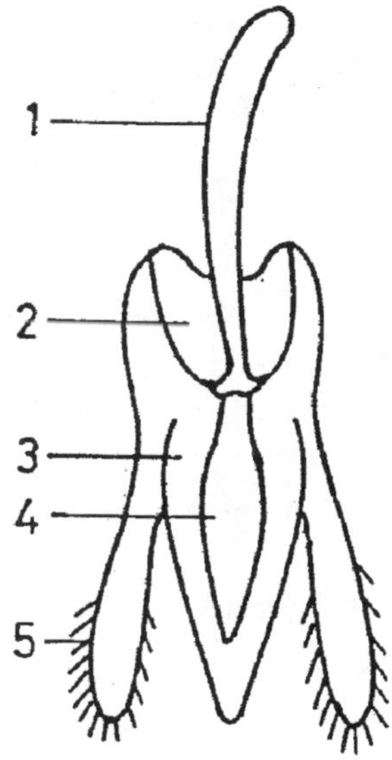

Plate XX

Fig. 110: Tegmen of Male

(1) Tegmen strut, (2) Basal piece, (3) Median lobe, (4) Groove, (5) Lateral lobe.

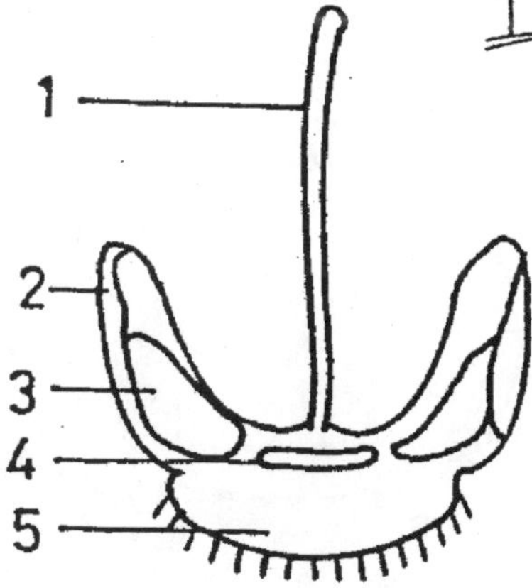

Plate XX

Fig. 111: Male Genitalia

(1) Apophysis, (2) 9th tergite, (3) 9th pleurite, (4) 9th sternite, (5) 10th tegite.

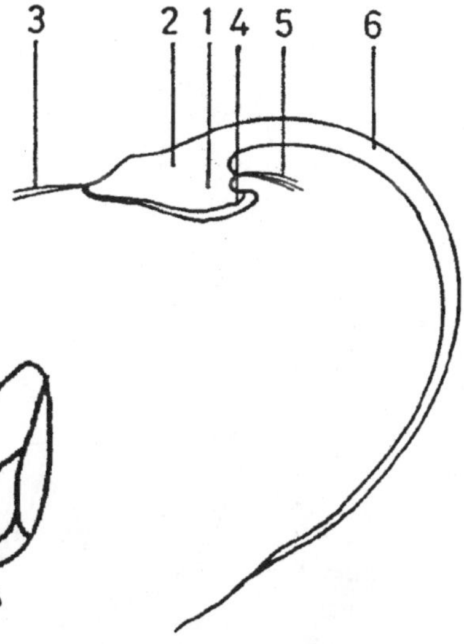

Plate XX

Fig. 112: Sipho

(1) Siphonal capsule, (2) Outer lobe of siphonal capsule, (3) Ejaculatory duct, (4) Inner lobe of siphonal capsule, (5) Trachea, (6) Siphonal tube.

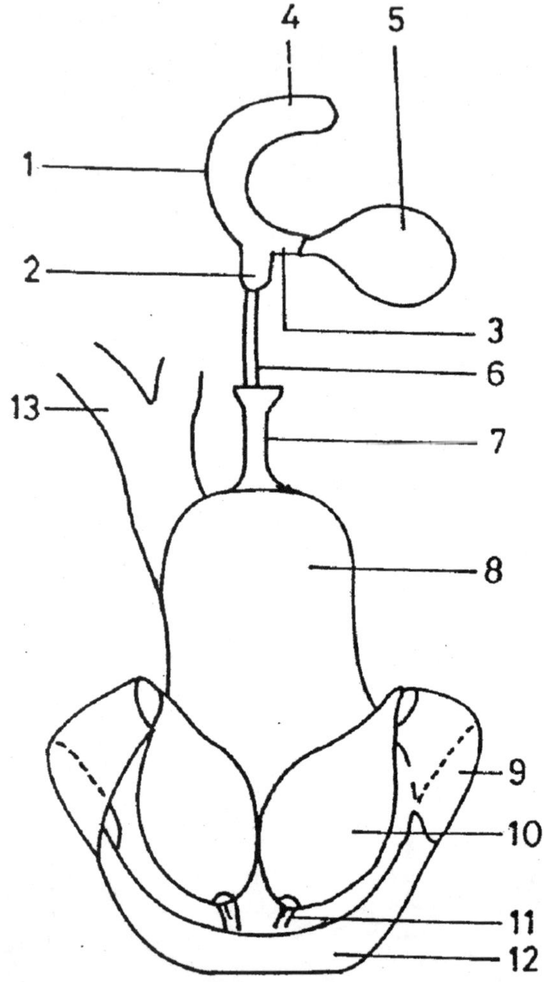

Plate XX

Fig. 113: Ovipositor and Sepermatheca

(1) Receptaculum seminis, (2) Nodulus, (3) Ramus, (4) Cornu, (5) Accessory gland, (6) Sperm duct, (7) Infundibulum, (8) Bursa copulatrix, (9) 9th pleurite, (10) Hemisternite, (11) Stylus, (12) 9th tergite, (13) Oviduct.

were collected from ecologically varying types of habitats, agro-ecosystems, orchards, forests, etc.

Mostly collection of lady bird beetles have been made at morning and evening throughout the year. Specimen were usually collected by sweeping or beating vegetation in cultivated and wooded areas and by hand picking.

Specimen were killed and then pinned and preserved in insect box (Fig. 11) and in 70% alcohol. Specimen preserved in alcohol were dissected to study the taxonomic characters.

To study the genitalia, specimen were dissected with the help of dissecting needle and forcep, after placing abdominal parts into hot solution of concentrated aqueous KOH for about 3 to 5 min. The prolonged exposure to KOH causes distortion of the genitalia. Genitalia were then mounted in DPX.

The larvae of predatory Coccinellids were also collected from field and reared in the laboratory for their adult emergence. Carabid beetles and praying mantids were collected from the different parts of Kolhapur district by visiting the fields (Fig. 12) frequently at evening and morning. The collected material was reared in the laboratory in insect glass cages (Fig. 2 & 3), glass troughs (Fig. 6) and plastic containers (Fig. 8) etc. The carabid beetles and praying mantids are preserved in insect boxes by pinning. After sorting of different groups and genera, each field collection was duly labelled with date of collection, locality and name of collector. Then antennae, labrum, mandible, maxilla, labium and hind leg were mounted in DPX on slides except praying mantids and large bettles, on card sheet.

With the help of monocular microscope morphological studies were carried out. Measurements were noted with eye piece micrometer and are in mm. All measurements calculated with the help of gaduated mechanical stage micrometer. The photographs of specimen were taken by using close up lenses while taxonomical parts were photographed microphotographically (Figs. 114-198). For identification and description of the coccinellid species the terms and terminology adopted here is the same as that of Sasaji (1968), Kapur (1970), Chapin (1974) and Sathe & Bhosale (2001).

The figures 203, 213, 219, 220 & 224 magnified to 1.5 times; the figures 202, 205, 210 & 216 magnified to 1.5 times; the figures 207,

208 & 222 magnified to 1.5 times; the figures 204 & 214 magnified to 2 times; figure 225 magnified to 2.5 times; the figure 221 magnified to 2.8 times; figures 114, 121, 129, 174, 193, 206 & 215 magnified to 3 times; figures 209 & 226 magnified to 3 times; figures 212, 218 & 223 magnified to 3.5 times; the figures 137, 143, 149, 157, 165, 175, 176, 177, 187 & 189 magnified to 4 times; the figures 173, 188, 190 to 192 magnified to 10 times; the figures 119, 125, 134, 182 & 192 magnified to 25 times; the figures 133, 140, 141, 146, 153, 171, 194 to 197 magnified to 40 times; the figures 118, 122, 154, 162 & 172 magnified to 50 times; the figures 130 to 132, 160, 168, 184 & 198 magnified to 60 times; the figures 123, 128, 138, 144, 150, 158, 166 & 185 magnified to 80 times; the figures 115, 116, 139, 142, 145, 152, 159, 161, 167, 169 & 194 magnified to 90 times; the figures 117, 120, 124, 135, 136, 147, 148, 155, 156, 163, 164, 170 & 183 magnified to 100 times and the figures 126 & 127 magnified to 120 times to its original size.

GENUS COCCINELLA LINNAEUS, 1758

The genus *Coccinella* is erected by Linnaeus in 1758 and is primarily holarctic (Dobzansky, 1931). Casey (1899, 1908) studied the North American species of *Coccinella* who recognized more than twice as many forms as separate species and many subspecies as did Leng (1903). Dobzansky (1931) described 11 species and many subspecies using male genitalia as the principal character. Brown (1962) studied 18 forms and reported 12 new species. The species described by other workers under this genus are *C. novemnotata* (Herbsh, 1793); *C. transversalis* (Fabricius, 1781); *C. nigrita, C. septempunctata, C. repanda* (Thumberg, 1781); *C. dimidia* (Hope, 1831); *C. basalis* (Redtenbacher, 1848); *C. ainu* (Lewis, 1896); *C. explanata, C. haseqawao* (Miyatake, 1963); *C. quinouepunctata* (Linnaeus), *C. transversoguttata quinoquenotata* (Felderman), *C. trifasciata perflexa* (Mulsant) and *C. transversoquattat richardsoni* (Brown) from the Peach orchards of Canada. *C. septumpunctata* is reported from Ukraine Yadechke (1954) as a typical species of cereal fields. From India, Kapur (1963) reported 3 new species namely, *C. tibetina, C. lama* and *C. nigroivattata*. He also added on the redescription of *C. transversalis* (Fabricius). Recently, Sathe & Bhosale (2001) described four new species of this genus, *C. madhuri, C. bhuimungi, C. yaminii* and *C. yogini*.

The genus is characterized by body broadly oval, convex; head black with two pale spots on frons; antenna slightly longer than frons; post antennal process narrow and long; mesosternum flat to the anterior margin; scutellum triangular, elytra yellow, orange or red with variable black patterns; post coxal line incomplete; presence of spurs on middle tibia.

Coccinella tabaci sp. nov.

(Plate - 21, Figs. 114 to 120)

Male

Body large, 6.3 mm long and 4.60 mm wide, oval, strongly convex on dorsal side; pronotum black, 1.55 mm long, glabrous; elytra dark yellowish brown, 5.1 mm long and 4.1 mm wide with three and half black spots; elytral anterior spot subrounded and slightly towards inner margin, ventral side black; antenna clavate, 0.60 mm long, brown, with dark black colonration on scape and last three segments; 9th tergite 'U' shaped, tegmen 0.90 mm long, siphonal tube 0.95 mm long.

Head

Black, 1.70 mm long and 2.02 mm wide, quadrilateral with rounded sides; setae on lateral sides; frons flat, weakly convex and rough, frontoclypeal suture not distinct; clypeus narrower, broadly expanded laterally and projected anteriorly; compound eyes placed antero-laterally, interocular distance 1.5 mm, surface rough and shiny; gular sutures indistinct, tentorial arms thin and parallel.

Antenna (Fig. 115)

11 segmented, 0.60 mm long; scape large, as wide as long, 2 mm long with fine setae on lateral sides; pedicel smaller than scape, somewhat quadrilateral, as wide as long with 2 long fine setae on one side; 1st segment longer than other segments of flagellum; flagellar last 3 segments large, 3 to 4 mm long with fine setae; 7th and 8th segments with single seta, 9th segment with 4 setae.

Flagellar formula

SL/W = 0.13, PL/W = 0.12, 2 L/W = 1, 5 L/W = 1.6, 7 L/W = 0.83, 8 L /W = 0, 5.9 L /W = 0 87.

Labrum

Somewhat quadrate, 0.20 mm long and 0.30 mm wide, lateral

borders rounded and in middle angulate, long setae on anterior border, labrum dark brown.

Mandible (Fig. 116)

Somewhat subtriangular, 0.28 mm long and 0.20 mm wide; apical tooth bifid with 0.05 mm depression from marginal line and 0.03 mm from apical tooth; basal tooth pointed, curved and triangular; membranous prostheca 0.19 mm, mandibles dark brown.

Maxilla

Cardo 0.23 mm long and 0.28 mm wide, as wide as stipe, stipe longer than wide, 0.42 mm long and 0.25 mm wide; galea 0.35 mm long, covered by thick, long setae; lacinia 0.31 mm long and covered by thick long setae; maxillary palpus 4 segmented, 1.35 mm long, 1st segment smaller than others, 0.10 mm long, terminal segment securiform, 0.69 mm long and 0.63 mm wide, apically truncate, maxilla dark brown.

Labium

Postmentum indistinct from gula; mentum quadrilateral, 0.23 mm long and 0.33 mm wide; prementuim sub-quadrate, 0.22 mm long and 0.35 mm wide, with fine hairs on outer border; labial palpus 3 segmented, 0.42 mm long; 1st segment 0.8 mm long, small, 2nd 0.12 mm long; 3rd 0.16 mm long, well developed and terminate roundly; labium dark brown.

Thorax

1.40 mm long and 3.75 mm wide; prothorax broader than long, convergent and anteriorly inclined, anterior margin broadly excavated and cover the head; pronotum black, 1.65 mm long and 3.14 mm wide, glabrous with yellowish spots on anterior corners, anterior corners angulate, posterior border covered by setae; middle coxal cavity narrowly opened outwardly; hind coxal cavity strongly transverse and entire metathorax dorsally concealed by elytra; metendosternite "Y" shaped with broad stalk and paired lateral arms of fulcra; scutellum black.

Elytra (Fig. 114)

5.1 mm long and 4.1 mm wide, heavily sclerotized, brownish yellow, glabrous, granulose, each elytron with three and half black spots; 1st post transverse band like with median elevation and not touching to inner and outer border of elytron; median spot also band like, irregular, not touching to outer border and fuses on suture with its counterpart; apical spot small, band like and fused with counterpart and forms subrounded spot and slightly forwards inner margin black.

Hind leg (Fig. 117)

Coxa 0.50 mm long and 0.12 mm wide, stout; trochanter 0.30 mm long and 0.10 mm wide, subtrapezoidal with straight inner margin, apical corner sharply denticulate; femur broad, 0.70 mm long and 0.30 mm wide, broader in middle and narrower towards the apex, fine setae on outer margin; tibia elongated, 0.65 mm long, 0.13 mm wide, narrow at base and broader at apex, covered by fine dense setae, setae more towards apex; tarsi 4 segmented 0.30 mm long and 0.19 mm wide, covered by setae on outer margin; elongated gradually broader towards apex and terminate into bifid claw; claw 0.05 mm long; curved, pointed and with basal tooth.

Abdomen

3.02 mm long, 4.45 mm wide, dorsum black, 8 tergite visible, smooth and without hairs; ventral side hairy, black, 6 sternites visible; lateral broader with setae; coxal line complete; femoral line angulate and terminate very close to antero-lateral side of 1st abdominal sternite, area around the femoral line brownish black, punctuate sparsely.

Male genitalia

9th tergite 'U' shaped, 0.43 mm long and 1.69 mm wide; 10th tergite strongly transverse, 1.02 mm wide; 9th pleurite attached to 9th tergite; 9th sternite small; apophysis elongate, 1.02 mm long, slender with slightly expanded at base.

Sipho (Fig. 120)

Siphonal capsule well developed 0.29 mm long and 0.16 mm wide, outer lobe quadrate, 0.20 mm long and 0.40 mm wide, inner lobe of capsule 0.10 mm long and 0.20 mm wide, immediately curved, gradually narrower toward apex, 0.12 mm wide; Siphonal tube 0.95 mm long.

Tegmen (Fig. 119)

Tegmen strut short, 0.90 mm long and 0.23 mm wide, slightly broad; lateral lobe, 0.0.30 mm long elongated; narrowed with dilated apex covered by dense setae; median lobe 0.50 mm long and 0.18

mm wide, flat with pointed triangular apex and deep groove on ventral side for sipho.

Female

Body larger than male 5.70 mm long and 4.80 mm wide, oval, similar like male except body size. *Female genitalia*: 9th pleurite and a pair of hemisternites together form ovipositor, hemisternites somewhat triangular with pair of stylus. *Receptaculum seminis* Spermathica (Fig.118): strongly 'U' shaped, broad at base and narrow at apex, markings on surface, cornu 0.22 mm, ramus 0.23 mm and nodulus 0.09 mm long.

Holotype

Male, India; Maharashtra, Kagal Coll., 6-IX-01, Vaishali. J. Patil on Tobacco (*Nicotiana tabacum* Linn.) antenna, mandible, hind leg, sipho, Tegmen, and spermatheca mounted on slides, labelled as above.

Paratype

Male-3, Female-3, Sex ratio (Male: Female) 1 : 1, data same as in holotype, Coll. 6-IX-01 V. J. Patil, India, Maharashtra, Kagal.

Prey

Aphid, *Myzus persicae* Sulz.

Prey Plant

Tobacco, *N. tabacum*.

***Coccinella tabaci* sp. nov. runs close to *C. transversalis* by having following characters.**

1. General body colour
2. Body size.
3. Labrum somewhat quadrate and mandibles.

   However, it differs from the above species by having following characters.

   1. Flagellar Formula -

      SL/W = 0.13, PL/W=0.12, 2L/W=I, 5LW=I.6, 7L/W=0.83, 8L/W = 0.5, 9L/W = 0.87.
   2. Mandibles slightly bifid and angularity curved.
   3. Mouth parts dark brown.

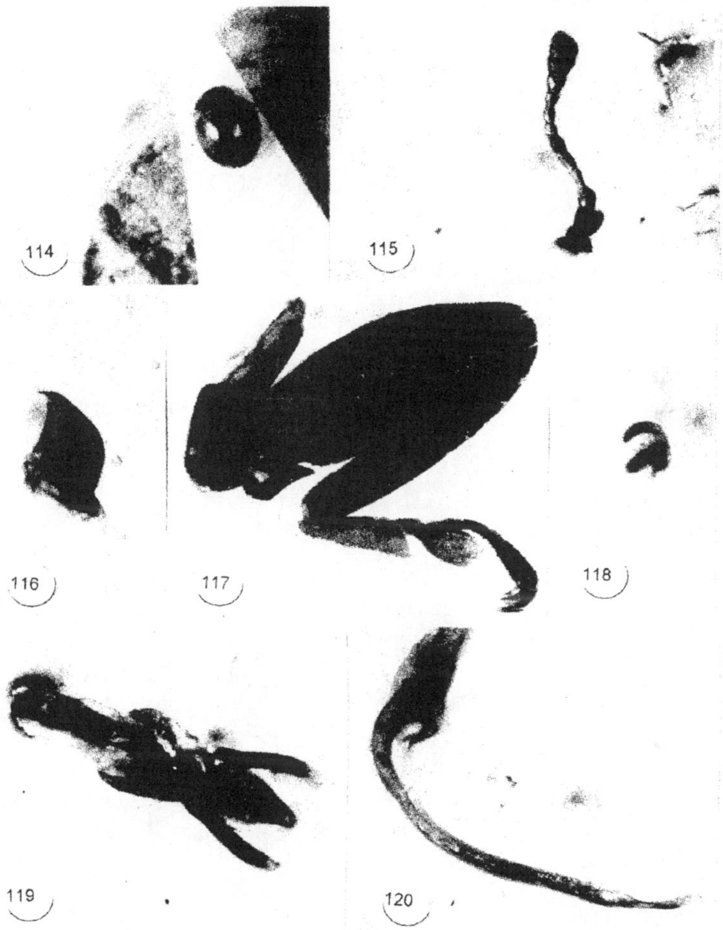

Plate XXI: ***Coccinella tabaci*** **sp. nov. (Figs. 114-120)**

Fig. 114: Adult, Fig. 115: Antenna, Fig. 116: Mandible, Fig. 117: Hind leg, Fig. 118: Spermathica, Fig. 119: Tegmen, Fig. 120: Sipho.

4. Elytral anterior spot sub-rounded and slightly towards inner margin.
5. Apex of lateral lobe of tegmen is densely hairy.
6. Antenna with fine two setae on lateral sides.
7. Spermathica strongly 'U' shaped.

GENUS *ILLEIS* MULSANT, 1850

Mulsant erected this genus for the first time in 1850. *Illeis* Mulsant, 1850, *Securipalp* 1026.

Egleis Mulsant, 1850. I.C.: 167.

Type species: *Coccinella cincta* Fabricius, 1798.

There are five species described under this genus namely, *Illeis cincta, I. bistigmosa* and *I. indica* and a subspecies *I. cincta amamiana* (Miyatake, 1959; Crotch, 1874; Korschefsky, 1932; Timberlake, 1943; Bielawski, 1961), *I. darbari* & *I. sathei* (Sathe & Bhosale, 2001). From India only two species have been reported (*I. darbari* & *I. Sathei*) by Sathe & Bhosale (2001) .

The genus is characterized by body medium sized, short, oval and weakly convex; frons narrow; eyes with a shallow postantennal emargination; antennae distinctly longer than head width and very thin; mandibles usually with multidenticulate apex, rarely with a bifid apex, terminal segment of maxillary palpus strongly transverse; scutellum transverse and triangular; poststernal process with parallel carinae; femoral line of 1 st abdominal sternum incomplete; tibiae slender and simple, without spurs, tarsal claws with a quadrate tooth at its base.

Illeis chilliei sp. nov.
(Plate 22, Figs. 121 to 128)

Male

Body short, 5.40 mm long and 4.20 mm wide, slightly elongated whitish yellow; pronotum 0.70 mm long and 1.10 mm wide, with 2 black spots on posterior margin; antenna 1.30 mm long; elytra 2.00 mm long and 1.40 mm wide; middle and apex of lateral lobes of tegmen covered by setae.

Head

Black 1.40 mm long and 1.90 mm wide, some what quadrate with slightly projected anterior side; frons pale, fronto clypeal suture indistinct; clypeus expanded in front of compound eyes; mouth parts yellowish brown; compound eyes black, laterally projected, interocular distance 0.30 mm, dorsum smooth; lateral borders covered by setae; ventrum smooth, black; gular sutures prominent and run parallel; tentorium with tentorial arms.

Antenna (Fig. 122)

11 segmented, 1.30 mm long, clavate and somewhat serrated; scape large, more longer than wide with six fine setae on inner side; pedicel as broad as long, somewhat quadrate with 2 fine setae on inner side; 1st, 2nd and 3rd segments longer than other segments each with one fine seta on inner side; 4th, 5th and 6th segments more or less similar, 5th segment with one long seta and 5 fine setae on inner anterolateral side; 7th, 8th and 9th segments larger than preceding segments and forms distinct club, 9th segment larger than others, narrow at base and apically broader with 6 long setae; scape and terminal segment yellowish with brown tinge.

Flagellar formula

1 L/W = 0.2 mm, 2 L/W = 1.75 mm, 4 L/W = 1.66 mm, 6 L/W = 2.00 mm, 7 L/W = 1.75 mm, 8 L/W = 1.2 mm, 9 L/W = 0.14 mm.

Labrum

Oval, yellowish, subrounded, lateral sides covered by setae elevations in middle.

Mandible (Fig. 124)

Somewhat triangular, 0.20 mm long and 0.14 mm wide; apical tooth bifid, reddish, sharply pointed with 0.03 mm depression from marginal line and 0.04 mm from apical tooth; basal tooth small, very slightly curved and abruptly pointed; membranous prostheca 0.6 mm.

Maxilla (Fig. 123)

Cardo 0.11 mm, as broad as long, 0.18 mm wide; stipe longer than width 0.15 mm long and 0.6 mm wide; galea large, hook like 0.25 mm long, with thick dense setae; lacinia 0.15 mm long, small, simple, leaf like with fine setae on inner side; maxillary palpus 4 segmented; basal segment small; 2nd segment narrow at base and wide, small, quadrate; terminal segment large, more transverse, capitulate, narrow at base and apically expanded.

Labium

Post mentum fused with gula; mentum broad, prementum triangular, apically broad and not covered by hairs; labial palpus 3 segmented, 1st segment very small, 2nd segment as long as broad, 3rd segment bluntly pointed.

Thorax

1.70 mm long and 2.50 mm wide; prothorax broader than long; Pronotum 1.00 mm long and 2.30 mm wide, quadrate, pale yellow, smooth, having 2 black spots and with rounded lateral borders; hypomerone slightly foveate; prosternum with parallel carinae not reaching to anterior border of sternum; anterior margin of mesosternum slightly convex; metasternum slightly triangular .

Elytra (Fig. 121)

2.79 mm long and 1.40 mm wide, pale yellowish without spots, finely punctuate in middle and densely punctuate on outer margin; when elytra on body, it shows black stripes; elytral epipleuron pale, without punctures, transversely broad and reaching to apex.

Hind leg (Fig. 126)

Coxa 0.70 mm long, 0.24 mm wide, shortly elongated and broad; trochanter somewhat triangular, cap like on femur with prominent denticulate process on inner side; femur elongated, 0.65 mm long and 0.11 mm wide, broad in middle, tibia not elongated, 0.16 mm long and 0.7 mm wide, cylindrical with tibial spines on all sides and tibial spurs absent; tarsi 3 segmented 0.23 mm long and 0.4 mm wide, 2nd segment triangular with setae on anterior border; terminal segment long, flattened, narrow at base and gradually broad towards apex and terminate into bifid claw; basal tooth pointed with subquadration.

Abdomen

2.30 mm long and 3.00 mm wide, dorsum yellowish; ventrum pubescent; 8th terga visible, broader at base and narrower towards apex, 6th sternite visible; lateral border covered by setae; coxal line on 1st abdominal sternite straight; femoral line incomplete and reaching to posterior. border of sternum (Fig. 125).

Male genitalia

9th tergite 0.15 mm long and 1.20 mm wide, 10th tergite strongly transverse, 0.12 mm long and 0.14 mm wide, covered by setae; apophysis 0.62 mm long, elongated, base not bifid.

Sipho (Fig. 128)

Sipho 1.45 mm long, siphonal capsule 0.7 mm long and 0.9 mm wide, well developed, outer lobe quadrate, inner lobe curved; siphonal tube immediately curved, 1.35 mm long uniform in width up to middle and at apex blunt.

Tegmen (Fig. 127)

Tegmen 0.73 mm long, tegmen strut 0.48 mm long, similar in width throughout length, with slightly broad and flat at apex; lateral lobes 0.58 mm long, narrow in middle and slightly broad, rounded at apex with dense long setae on middle apex; median lobe 0.28 mm long, narrow towards apex and shows knife like but not with hook like apex.

Female

Unknown.

Holotype

Male, India: Maharashtra, Kolhapur Coll. 10-IX-02, Vaishali J. Patil, on Chilli (*Capsicum annuum* L.) antenna, mandible, maxilla, 1st abdominal sternite, hind leg, sipho and tegmen on slides labelled as above.

Prey

Aphid, *Aphis gossypii* Clover.

Prey Plant

Chilli (*Capsicum annuum* L.)

Discussion

Illeis chillie sp.nov. runs close to *I. darbari* (Sathe & Bhosale, 2001) by pronotal spots and general features of antenna. But it differs from above species by following characters.

1. Flagellar formula:

   1 L/W = 0.2 mm, 2 L/W = 1.75 mm, 4 L/W = 1.66 mm, 6 L/W = 2.00 mm, 7 L/W = 1.75 mm, 8 L/W = 1.2 mm, 9 L/W = 0.14 mm.

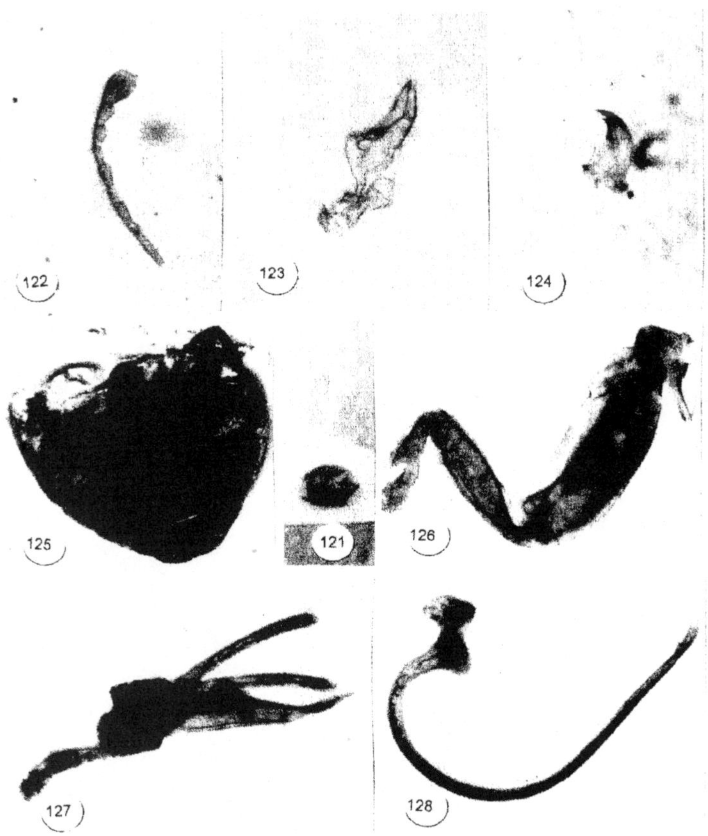

Plate XXII: *Illeis chilliei* sp. nov. (Figs. 121-128)

Fig. 121: Adult, Fig. 122: Antenna, Fig. 123: Maxilla, Fig. 124: Mandible, Fig. 125: 1st Abdominal Sternite, Fig. 126: Hind leg, Fig. 127: Tegmen, Fig. 128: Sipho.

2. Sipho-Siphonal tube immediately curved and not so blunt as in *I. darbari*.
3. Tegmen-knife like but not with hook like apex.
4. Mandibles dark reddish brown, basal tooth slightly curved and abruptly pointed.

Illeis gawari sp. nov.
(Plate 23, Figs. 129 to 136)

Male

Body oval, 4.65 mm long and 3.4 mm wide, dorsum convex; pronotum 1.30 mm long and 2.70 mm wide; antenna 0.75 mm long; elytra 3.75 mm long and 1.73 mm wide, with one black spot, yellowish; sipho 3.32 mm long.

Head

1.07 mm long and 1.65 mm wide; subrounded; pubescent dorsum; frons yellowish, indistinct fronto clypeal-suture; slightly broad clypeus, clypeus expanded in front of compound eyes; eye surface glabrous, interocular distance 0.60 mm; slightly distinct gular sutures.

Antenna (Fig. 130)

11 segmented including scape, pedicel and flagellum; short, 0.75 mm long, thin, serrated; scape small; pedical as long as wide, elongated; 1st & 2nd segments smaller than other segments by width, 5th segment quadrate with rounded apex; 6th segment broader than 5th segment and smaller than 7th segment, 7th segment rectangular, wider than long. 8th segment large, yellowish, with fine setae; 9th segment, oval, apex slightly narrow.

Flagellar formula

SL/W = 1.20, PL/W = 1.5, 2L/W =1.1, 4L/W = 1.8, 7L/W = 1.4, 8L/W = 1.2, 9L/W = 1.22.

Labrum

0.32 mm long and 0.41 mm wide, plate like rounded, yellowish brown, rounded anteriorly and posteriorly, oval transversely.

Mandible (Fig. 132)

Elongately triangular, 0.30 mm long and 0.20 mm wide, dark reddish brown, apical tooth bifid, not so pointed, curved, centrally broad, basal portion somewhat curved and dark brown, depression 0.16 mm from marginal line and 0.8 mm from apical tooth; basal tooth present, reddish brown, pointed, curved; membranous prostheca 0.19 mm, dark brown.

Maxilla (Fig. 131)

Cardo smaller than stipe, 0.16 mm long and 0.18 mm wide; stipe longer than width, 0.25 mm long; galea large, 0.17 mm long; simple lacinia, 0.7 mm long; maxillary palpus 4 segmented, 0.98 mm long, basal segment narrow at base and broad at apex; 2nd segment quadrate with rounded corners, dark reddish brown, 0.29 mm long; terminal segment 0.34 mm long and 0.93 mm wide, somewhat triangular, expanded at apex and smaller at base; rounded corners.

Labium

Gula fused with poststernum; squarish mentum, 0.11 mm long and 0.14 mm wide, covered by fine hairs apically, broad; 3 segmented labial palpus, 0.33 mm long, basal segment small, 2nd and 3rd segments more or less similar, 3rd segment broad at base with narrow and blunt pointed apex.

Thorax

1.18 mm long and 3.6 mm wide; yellowish pronotum, 1.22 mm long and 2.59 mm wide, with two black spots on posterior margin, spots small and sub-rounded, rounded lateral corners; straight poststernal carinae, without posterior end; metandostemite with pair of fulcra; scutellum triangular.

Elytra (Fig. 129)

Light, yellowish, 3.75 mm long and 1.73 mm wide having one small black spot on elytra at inner side; elytra anteriorly rounded, outer border inclined towards apex and not pointed; elytra, convex, oval.

Hind leg (Fig. 134)

Coxa small, rounded, 1.8 mm long, yellowish brown, stout; trochanter 0.28 mm long and 0.21 mm wide, yellowish elongated; femur 1.60 mm long and 0.70 mm broad, curved at outer side, outer side brown and inner side yellowish, small, somewhat straight, narrow at anterior end; tibia 1.55 mm long and 0.20 mm wide, brown, elongated, narrow and broad at anteriorly, smaller in size at posterior end, middle broad; with dark blackish spot, tarsi, yellowish brown, 0.56 mm long, apex broad and straight, terminate into bifid claw; claw pointed, 0.15 mm long, curved basal tooth, brown.

Abdomen

2.41 mm long and 2.52 mm wide, yellowish 6th sternites visible, lateral border yellowish, coxal line straight; femoral line incomplete, area rounded, femoral line smooth, transverse band dark brown, last segment narrow, dark brown (Fig. 133).

Male genitalia

9th tergite 'V' shaped, 10th tergite transverse and covered by setae; apophysis elongated with rounded apex.

Sipho (Fig. 136)

3.32 mm long, siphonal capsule not well developed, small, 0.12 mm long and 0.35 mm wide; outer lobe 0.10 mm long and 0.19 mm wide, smaller than outer lobe and slightly curved; siphonal tube 3.12 mm long and 0.06 mm wide, curved, gradually width decreased and apex pointed, yellowish.

Tegmen (Fig. 135)

Tegmen strut 0.40 mm long and 0.6 mm wide, broad at posterior end, lateral lobes narrow at apex with long setae; median lobe knife like apex; yellowish brown; lateral lobes longer than median lobe.

Female

Unknown.

Holotype

Male, India: Maharashtra, Hatkanangale, Coll. 21-IX-2000, Vaishali J. Patil, on Gawar (*Cyamposis tetragonoloba* L.), antenna, mandible, maxilla, 1st abdominal sternite, hind leg, sipho and tegmen on slides labelled as above.

Paratype

Male, 1, data same as in holotype, Coll. 21-IX-2000, Vaishali J. Patil, on Gawar.

Prey

Aphid (*Aphis gossypii*) Glover.

Preyplant

Gawar (*C. tetragonoloba*).

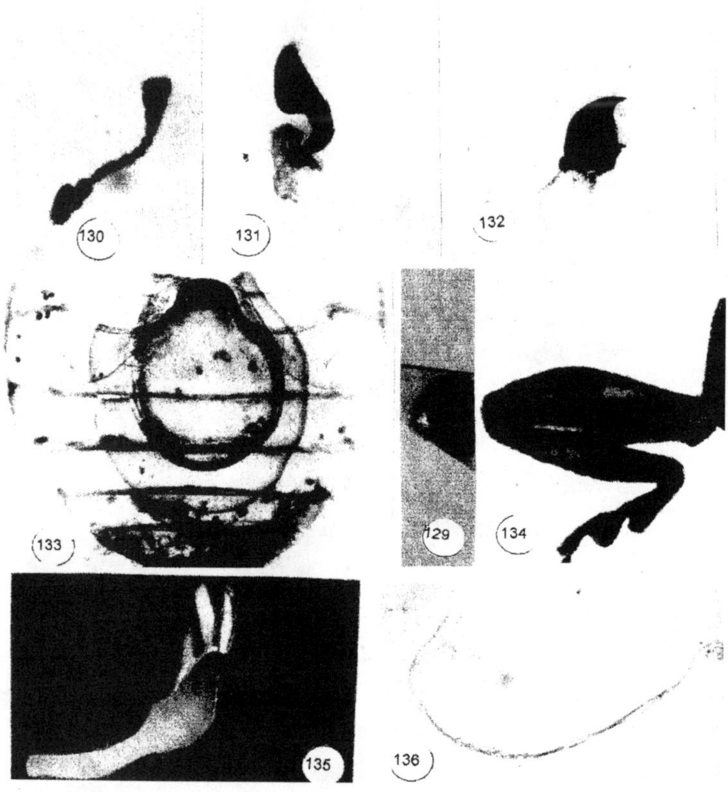

Plate XXIII: *Illeis gawari* sp. nov. (Figs. 129-136)

Fig. 129: Adult, Fig. 130: Antenna, Fig. 131: Maxilla, Fig. 132: Mandible, Fig. 133: 1st Abdominal Sternite, Fig. 134: Hind leg, Fig. 135: Tegmen, Fig. 136: Sipho.

Discussion

This species runs close to *I. darabari* S & B (Sathe & Bhosale, 2001) by following characters, pronotal spots, labium, legs, abdominal sternite, last terga of male. However, it differs from the above species by having the following characters:

1. Labrum oval transversely, rounded anteriorly and posteriorly.
2. Mandible elongately triangular, with basal tooth.
3. Elytra light yellowish with single black spot centrally.
4. Siphonal capsule not well developed, siphonal tube curved, gradually width decreased.
5. Flagellar Formula:

   SL/W = 1.20, PL/W = 1.5, 2L/W = 1.1, 4L/W = 1.8, 7L/W = 1.4, 8L/W = 1.2, 9L/W = 1.22.

Illeis therioaphis sp. nov.

(Plate 24, Fig. 137 to 142)

Female

Body short oval, light yellowish brown, 4.85 mm long and 3.60 mm wide, dorsum weakly convex; pronotum 1.27 mm long and 2.63 mm wide; antenna 0.80 mm long; elytra 3.83 mm long and 1.80 mm wide, coarsely punctured and without spots; spermathica 0.28 mm long and 0.8 mm wide, 'C' shaped.

Head

1.14 mm long and 1.40 mm wide, subrounded; dorsum pubescent; posterior border coated by hairs; frons brown and rather narrow, finely pubescent, frontoclypeal suture indistinct; clypeus slightly broad and expanded infront of compound eyes; glabrous surface of eye, interocular distance 0.52 mm; eyes large, slightly distinct gular sutures; tentorium with parallel arms.

Antenna (Fig. 138)

11 segmented, 0.80 mm long, serrated and thin; longer than head; scape longer than width, large with long setae; pedicel sub quadrate, as long as wide; 1st segment with 2 setae, longer than wide; 2nd and 3rd segments equal; 7th and 8th segments distinctly

longer than wide; 9th segment larger than other segments, elongated oval and covered by long setae apically.

Flagellar Formula

5L/W = 1.29, PL/W = 1.8, 2L/W = 1.25,4 L/W = 1.3, 7 L/W = 1.6, 8 L/W = 1.2, 9L/W = 1.28.

Labrum (Fig. 142)

0.27 mm long and 0.47 mm wide, somewhat rounded, elongate laterally, anterior corners rounded, reddish brown.

Mandible (Fig. 139)

Elongately triangular, 0.28 mm long and 0.24 mm wide; apical tooth bifid, dark reddish brown, mostly pointed, curved, with 0.18 mm depression from marginal line and 0.5 mm depression from apical tooth; basal tooth present, small, reddish brown; membranous prostheca 0.20 mm, yellowish.

Maxilla

Cardo wider than stipe, 0.19 mm long and 0.21 mm wide; stipe more longer than width, 0.29 mm long; galea large, 0.19 mm long, spatulate and covered by thick long setae; simple lacinia, 0.8 mm long, leaf like with setae; 4 segmented maxillary palpus, strongly transverse, triangular, 0.90 mm long, small basal segment; 2nd segment longer than 3rd, 0.30 mm long; small 3rd segment 0.12 mm long, as wide as long, quadrate; large terminal segment 0.30 mm long and 0.82 mm wide, inner and outer borders expanded, transversely capitulate, with rough hairy surface.

Labium

Gula fused with postmentum; squarish mentum, 0.10 mm long and 0.17 mm wide, numerous fine hairs covered by apically and broad; 3 segmented labial palpus 0.35 mm long, small basal segment, 2nd and 3rd segments more or less similar, 3rd segment broad at base with blunt and narrow pointed apex.

Thorax

1.39 mm long and 3.10 mm wide; pronotum anterior margin weakly and roundly concave; brownish, 1.26 mm long and 2.63 mm wide with two black spots on posterior margin, spots large and rounded, lateral corners somewhat rounded, very fine punctured surface area; somewhat straight poststernal carinae, excluding posterior end; margin of mesosternum accurately emerginate anteriorly, 'Y' shaped metendosternite with pair of fulcra; transverse triangular scutellum.

Elytra (Fig. 137)

Brownish, 3.83 mm long and 1.80 mm wide without spots, rounded anterior corners, outer border inclined towards apex and pointed; elytral epipluron as wide as body width.

Hind Leg (Fig. 141)

Coxa 1.12 mm long and 0.38 mm wide, stout reddish brown, cylindrical, elongated, broad at posterior end; trochanter 0.32 mm long and 0.26 mm wide; yellowish brown, with denticulate process on inner side; femur 1.68 mm long and 0.59 mm wide, large, yellowish dark brown at anterior end, broad at middle, narrow at end; tibia 1.69 mm long and 0.25 mm wide, slender and simple without spurs, cylindrical, smaller at anterior end, dark brown, coating of hairs on inner side; tarsi dark brown, 0.65 mm long, apex broad, terminate into bifid claw; pointed claws 0.18 mm long, immediately curved with basal tooth, tarsal claws with a quadrate tooth at its base.

Abdomen

2.32 mm long and 2.68 mm wide, brownish, 6th sternite visible, lateral borders covered by dense hairs; straight coxal line; incomplete femoral line on 1st abdominal sternite, smooth area around femoral line.

Female Genitalia

9th tergite, a pair of 9th pleurite and a pair of hemisternites together form ovipositor; hemisternite somewhat oval, with elongated stylus.

Spermathica (Fig. 140)

'C' shaped, 0.28 mm long, 0.8 mm wide, whitish yellow, depression on lateral side, narrow towards apex and broad at base.

Male

Unknown.

Holotype

Female, India: Maharashtra, Kagal, Coil. 20-VII-2001, Vaishali J. Patil, on soyabean (*Glycine soja* Linn.) antenna, labrum, mandible, hind leg, spermathica on slide labelled as above.

Prey

Aphid, *Therioaphis sexmaculata* Buck.

Prey Plant

Soyabean, *Glycine soja* Linn.

Discussion

This species runs close to *I. koebeli* (Timberlake) (Kapur, 1966) by following characters.

1. Body medium sized, short oval.
2. Antenna yellowish brown.
3. Mouth parts yellowish brown.
4. Pronotum with two rounded black spots.

However, it differs from the above species by having following characters:

1. Body, colour, light yellowish brown.
2. Frons brown and narrow.
3. Mandible elongately triangular with bifid apical and basal tooth, only tooth dark brown.
4. Labrum elongate laterally, convex laterally.
5. Spermathica whitish 'C' shaped, broader at one end and narrow at another end.
6. Flagellar Formula:
   SL/W = 1.29, PL/W = 1.8, 2L/W = 1.25, 4L/W = 1.3, 7L/W = 1.6, 8L/W = 1.2, 9L/W = 1.28.

GENUS MENOCHILUS TIMBERLAKE, 1943

Menochilus: Timberlake, 1943, *Hawaii plant*. Rec 47(1): 40

Chilomenes: Mulsant, 1850, *Trim Spec*. securiplap: 429

Chilomenes: Grot.ch, 1874. *Rev. Cocci*: 179 (Pars)

Type species: Coccinella 6 *maculata Fabricius*, 1781.

The genus *Menochilus* is represented by Timberlake in 1943 which contains four species, *M. sexmaculatus* described by Fabricius in 1781; *M. tuljapurensis, M. shivajensis* and *M. Kharipi* described by Sathe & Bhosale in 2001. In the present text one more new species is

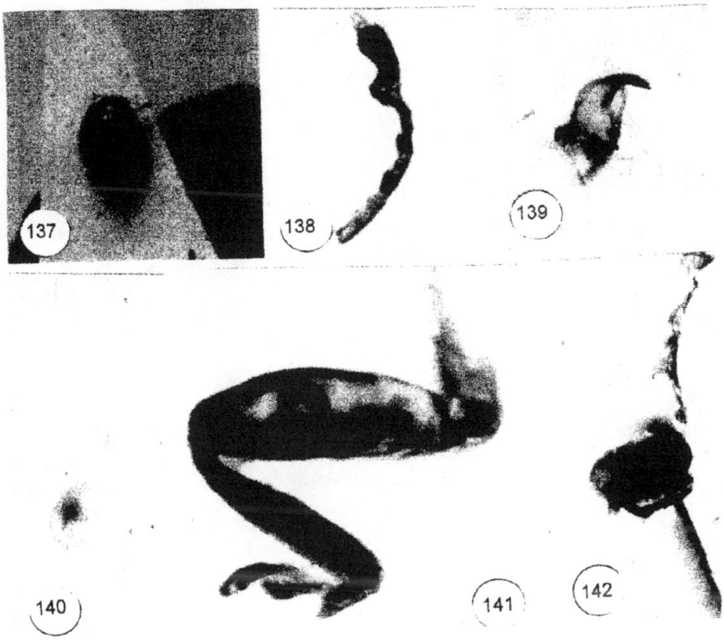

Plate XXIV: *Illeis therioaphis* sp. nov. (Figs. 137-142)

Fig. 137: Adult, Fig. 138: Antenna, Fig. 139: Mandible, Fig. 140: Spermathica, Fig. 141: Hind leg, Fig. 142: Labrum.

added to this genus. The genus is characterized by body rounded, slightly longer than wide and medium sized; prosternum with two carinae; anterior margin of mesosternum bisinuate and weakly emerginate in middle; elytral base broader, external margin of elytra, not expanded but inclined; 1st abdominal sternum with an oblique lateral line; middle and hind tibia with two spurs.

Menochilus vaishalii sp. nov.

(Plate 25, Fig. 143 to 148).

Male

Body 5.85 mm long and 4.3 mm wide, elongated, flat, longer than wide, orange brown; pronotum 1.10 mm long and 2.45 mm wide, antenna 0.72 mm long; elytra 4.9 mm long and 2.15 mm wide, with 3 black spots; sipho 2.85 mm long.

Head

1.36 mm long and 1.73 mm wide, rounded yellowish frons, fronto clypeal suture rough, clypeus yellow, anteriorly projected; eyes black, laterally projected, shiny; interocular distance 0.80 mm.

Antenna (Fig. 144)

11 segmented, 0.72 mm long, clavate; scape more broader than long, pedicel elongated, small; 1st to 4th segments equal; 5th segment somewhat rectangular, 6th segment longer than 5th segment terminal segments, 7th, 8th & 9th form compact club; terminal segment smaller than 8th segment, narrow, brownish.

Flagellar Formula

SL/W = 1, PL/W = 0.6, 1L/W = 1.4, 5;/W = 1, 7L/W = 0.75, 8L/W = 0.6, 9L/W = 1.

Labrum

0.26 mm long an 0.52 mm wide, quadrate, lateral borders, rounded, yellowish.

Mandible (Fig. 145)

0.28 mm long and 0.45 mm wide, apical tooth not bifid, basal tooth pointed, dark brown, transversely parallel to prostheca, membranous prostheca 0.16 mm long.

Maxilla

Cardo 0.13 mm long and 0.12 mm wide, quadrate; stipe 0.20 mm long and 0.17 mm wide, gradually narrowed towards apex; galea 0.18 mm long with coating of setae; lacinia 0.20 mm long, simple, leaf like with setae; maxillary palpus 4 segmented, 0.86 mm long, small basal segment, 0.09 mm long, 2nd segment 0.17 mm long and 0.07 mm wide, 3rd segment 0.19 mm long terminal segment large, 0.39 mm long.

Labium

Postmentum and gula fused; mentum 0.21 mm long and 0.16 mm wide, prementum quadrate, 0.24 mm long and wide, anteriorly covered by numerous hairs; labial palpus 3 segmented, 0.26 mm long, basal segment small, 0.04 mm long; 2nd segment 0.12 mm long and 0.09 mm wide; terminal segment 0.13 mm long and 0.06 mm wide with rounded apex.

Thorax

1.65 mm long and 2.39 mm wide, prothorax broad; pronotum 1.22 mm long and 2.45 mm wide, anterior margin rounded, lateral borders rounded, posterior border black; metendosternite with paired lateral arms, scutellum black.

Elytra (Fig. 143)

4.9 mm long and 2.15 mm wide with 3 black spots, flat, orange brown, depressed in middle, longer than wide, inner border black, elytra tesaceous.

Hind Leg (Fig. 146)

Coxa 0.86 mm long and 0.23 mm wide, rounded yellowish brown; trochanter 0.16 mm long and 0.31 mm wide, somewhat rounded, yellowish; femur 1.38 mm long and 0.43 mm wide elongated, curved at laterally posteriorly, yellowish brown lateral side, posteriorly dark brown; tibia 1.41 mm long and 0.30 mm wide with tibial spines, yellowish; elongated, outer border dark brown, narrow at base and broad at apex; tarsi 3 segmented somewhat triangular, yellowish; last segment terminated into bifid claw; claw 0.16 mm long with basal tooth.

Abdomen

2.76 mm long and 3.42 mm wide, dorsum black, ventral side hairy; 6th sternite visible; coxal line complete; complete femoral line with angulate structure.

Male Genitalia

9th tergite 0.21 mm long and 1.13 mm wide, "U" shaped; 10th tergite 0.12 mm long and 0.80 mm wide and strongly transverse with setae; apophysis 1.02 mm long with rounded apex.

Sipho (Fig. 148)

2.85 mm long; siphonal capsule not well developed, 0.11 mm

long and 0.38 mm wide, outer lobe 0.13 mm long and 0.21 mm wide, innerlobe short; siphonal tube 2.65 mm long, not immediately curved, very thin tube, whitish yellowish, more longer and pointed.

Tegmen (Fig. 147)

Lateral lobes longer than median lobe, 0.71 mm long, narrow apex and covered by setae, yellowish; median lobe with pointed apex, flat, broad at base; tegmen strut 1.18 mm long, thin, yellowish, elongated.

Holotype

Male, India: Maharashtra, Kolhapur Coll. 25-IX-2001, Vaishali J. Patil, on Jowar (*Sorghum vulgare* Pers.); antenna, mandible, sipho, tegmen and hind leg on slides labelled as above.

Paratype

Male 13, data same as in holotype Coll. from 25-IX-2001 to 12-VIII-2002, Vaishali J. Patil on Jowar (*S. vulgare*); Maize (*Zea mays* Linn.) and hybrid jowar CSH-9, India: Maharashtra, Kolhapur.

Prey

Aphid, *Rhophalosiphum maidis* (Fitch).

Prey Plant

Jowar (*S. vulgare*), Maize (*Z. mays*), Hybrid Jowar (CSH-9).

Discussion

This species runs close to *M. shivajensis* (Sathe & Bhosale, 2001) by following character:

1. Body colour and general appearance of elytral spots.
2. Elytra testacious.
3. Labrum.

However, it differs from above species by having following characters:

1. Mandible immediately triangular with a very prominent basal tooth and blunt apical tooth, reddish brown.
2. Sipho, siphonal capsule not well developed, siphonal tube long, thin with thread like terminal end, siphonal tube not immediately curved.

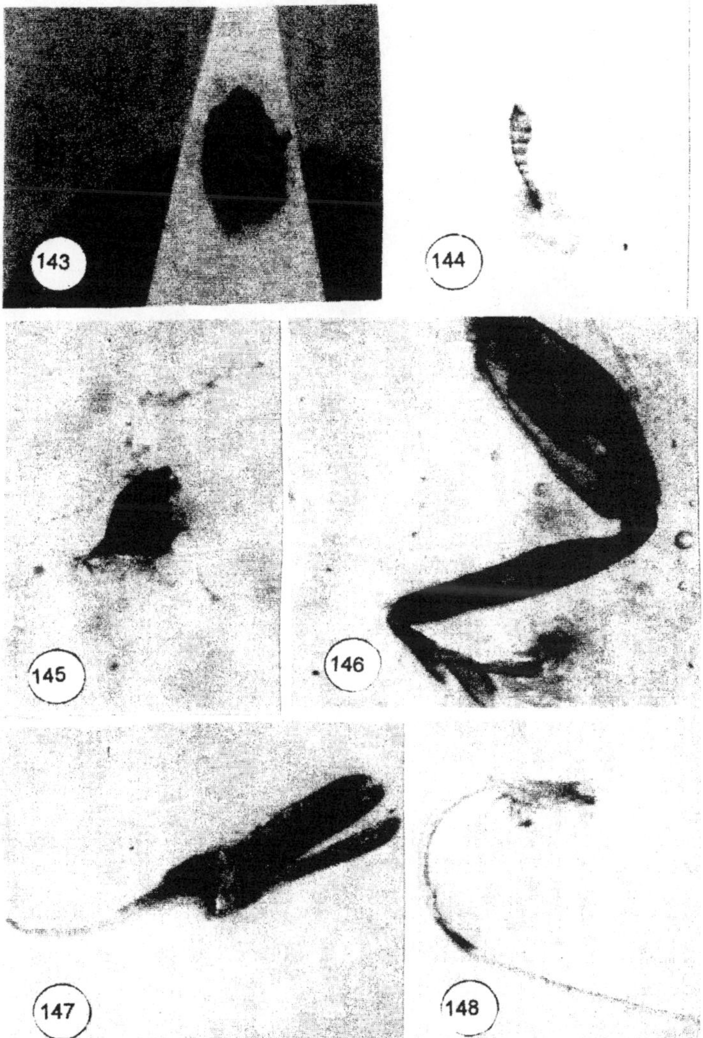

Plate XXV: ***Menochilus vaishalii*** **sp. nov. (Figs. 143-148)**

Fig. 143: Adult, Fig. 144: Antenna, Fig. 145: Mandible, Fig. 146: Hind leg, Fig. 147: Tegmen, Fig. 148: Sipho.

3. Tegmeninal median lobe with pointed apex, flat, yellowish.
4) Flagellar Formula:
   SL/W = 1, PL/W = 0.6, 1L/W = 1.4, 5L/W = 1, 7L/W = 0.75, 8L/W = 0.6, 9L/W = 1.

GENUS *MICRASPIS* DEJEAN, 1835

Micraspis Dejean, 1835, *Cat. Col.*, 459.

Vernia Mulsant, 1850, *Spec. Trim. Securipalp.*, 343.

Alesia Mulsant, 1850, 1. C: 343.

Type Species: *Coccinella striata* Fabricius, 1792.

From India four species have been reported (Bhaskar, 1992). In the present study two new species of *Micraspis* have been described from agro-ecosystem in Maharshtra.

The genus is characterized by following characters: body oval, dorsum convex, glabrous and medium sized; antenna comparatively short, slightly longer than frons, antennal club distinct; 9th segment transverse, 10th nearly twice as wide as longer than the preceding segment with a partly truncate apex and basal segment elongate; post antennal process of frons narrow and long; anterior margin of clypeus nearly straight; anterior margin of pronotum rather shallowly excavated, anterior corners angulate; lateral sides of pronotum without a depression along the margin and reflexed; prothorasic hypomeron very shallowly depressed; prosternal process with parallel carinae; anterior margin of mesosternum bisinuate, shallowly emarginate; scutellum triangular, small in size and narrower than pronotum; elytral base slightly but distinctly broader than pronotum; external side of elytra not broadly expanded but rather narrowly reflexed with a very narrow groove along the margin; elytral epipleuron as wide as body width, horizontal in inner part and inclined below in outer 1/3; inner carnia of epipleuron not reaching the apex but disappears at preapical segment of abdomen, 1st abdominal sternum with an incomplete femoral line; middle and hind tibiae with two spurs; claw with a basal tooth.

Micraspis gossypii sp. nov.
(Plate 26, Figs. 149 to 156)

Male

Body shortly oval and quadrate, 6.8 mm long and 4.5 mm wide, convex on dorsal side; 1.80 mm long and 4.20 mm wide pronotum, without spot and glabrous, antenna 11 segmented, 0.60 mm long and 0.5 mm wide; mouth parts yellowish brown; elytra 5.7 mm long and 2.25 mm wide with two black spots; sipho 1.50 mm long.

Head

1.65 mm long, 2.50 mm wide, yellowish brown, quadrate with rounded lateral sides; frons narrow and long with 2 small rounded black spots on posterior side; frontoclypeal suture not distinct; clyepus nearly straight, anterior margin expanded in front of compound eyes; eyes black, weakly projected laterally, 1.8 mm interocular distance; indistinct gular sutures and broadly separated; tentorium with very thin parallel arms.

Antenna (Fig. 150)

11 segmented, 0.60 mm long, short clavate; scape elongated posteriorly, curved to inner side; pedicel smaller than scape and longer than broad; 1st segment more longer than other segments, elongated; 2nd to 6th segments more or less quadrilateral and equal; 5th to 8th segments show prominent anterolateral corners; 9th segment transverse, broader than long; 9th segment larger, apically truncate with few fine setae.

Flagellar Formula

SL/W = 1.4, PL/W = 0.8, 2L/W = 1.25, 4L/W = 1.5, 7L/W = 2.3, 8L/W = 0.62, 9L/W = 1.14.

Labrum (Fig. 152)

0.15 mm long and 0.30 mm wide, quadrate, anterior corners rounded and covered with fine setae, yellowish and two dark brown spots at posterior corner.

Mandible (Fig. 151)

0.35 mm long and 0.18 mm wide, triangular with rounded base at corners; bifid apical tooth, dark reddish brown with 0.01 mm depression from marginal line; prominent, dark reddish brown coloured basal tooth; lateral borders dark brown, slightly curved; 0.22 mm long membranous prostheca.

Maxilla

Cardo as long as wide, 0.19 mm; stipe 0.34 mm long and 0.16 mm wide, stipe with broad base; 0.9 mm long, galea with hook like thick setae; 0.75 mm long, 4 segmented maxillary palpus; 1st segment 0.04 mm long, small; 2nd segment 0.12 mm long; 3rd segment 0.10 mm, small; terminal segment 0.45 mm long and 0.38 mm wide, large and securiform.

Labium

Gala fused with postmentum 0.26 mm long and wide mentum, flat; 0.21 mm long and 0.34 mm wide prementum with straight outer border, rounded anterior corners covered by fine setae; 3 segmented labial palpus, 0.35 mm long small basal segment longer than apical segment 0.12 mm long, elongated, tapering and bluntly rounded apical segment.

Thorax

2.8 mm long and 3.93 mm wide; prothorax broader than long; pronotum 1.80 mm long and 4.20 mm wide, glabrous and with 2 black spots, strongly convex, posterior border strongly arcuate and marginate lateral corners; slightly depressed hypomeron; prosternum with carinae, distinctly separated apically; mesosternum bisinuate, shallowly emarginate in the middle of anterior side; yellowish brown, triangular, slightly broader scutellum.

Elytra (Fig. 149)

5.7 mm long and 2.25 mm wide, brownish, glabrous with fine punctures; two black spots on each elytron at anterior side; suture not black; base broader and lateral borders not expanded, elytral epipuron as wide as body width.

Hind Leg (Fig. 154)

Brownish; coxa 0.24 mm long and 0.72 mm wide, cylindrical; trochanter 0.25 mm long and 0.18 mm wide, small, somewhat squarish and without denticulate process on inner side; femur 0.96 mm long and 0.33 mm wide, broad, elongated and broader in middle at outer margin; tibia 0.98 mm long and 0.15 mm wide, elongated and narrow at base and gradually broader towards apex with two tibial spurs; 4 segmented tarsi 0.30 mm long and 0.12 mm wide; 1st segment somewhat conical, 0.40 mm long and 0.10 mm wide, apex broad, setae covered by inner and lateral sides; 2nd segment 0.35 mm long and 0.9 mm wide, end rounded and covered by fine setae;

3rd segment small 0.06 mm long and 0.05 mm wide; terminal segment 0.38 mm long and 0.09 mm wide gradually broad towards apex, terminate into bifid claw.

Abdomen

2.35 mm long and 3.31 mm wide, brownish dorsum; visible sternite; hairy lateral borders; coxal line straight and complete abdominal sternite (Fig. 153), shows incomplete femoral line, angulate, area around femoral line hairy.

Male Genitalia

9th tergite 'V' shaped; 10th tergite transverse with coating of setae; apophysis elongate and uniform in width.

Sipho (Fig. 156)

1.50 mm long; siphonal capsule well developed, 0.25 mm long and 0.13 mm wide, well developed; siphonal tube immediately curved, uniform in width, dark reddish brown colour, with narrow apex, apex smaller; outer lobe 0.10 mm long and 0.20 mm wide; inner lobe 0.12 mm long and 0.10 mm wide.

Tegmen (Fig. 155)

Tegmen strut 0.45 mm long and 0.16 mm wide, broad at apex and reddish brown, elongated; lateral lobes 0.60 mm long and 0.6 mm wide, longer than median lobe with flat apex covered by dense setae; median lobe short than lateral lobes, 0.50 mm long and 0.7 mm wide.

Female

Unknown.

Holotype

Male, India: Maharashtra, Kolhapur, Coll 10-XII-2000, Vaishali J. Patil, on Cotton (*Gossypium hirsutum* Linn.), antenna, labrum, mandible, hind leg, sipho and tegmen on slides labelled as above.

Paratype

Male, 2, data same as in holotype Coll 10-XII-2000, Vaishali J. Patil, India: Maharashtra, Kolhapur.

Prey

Aphid *Aphis gossypii*, *A. craccivora*.

Prey Plant

Cotton *G. hirsutum*.

Discussion

This species runs close to *Micraspis discolour* (Fabricius) by having following characters:

1. General antennal shape.
2. Maxillary palpus.
3. Leg yellowish brown.
4. Mid and hind coxae black.
5. Coxal line straight and complete.
6. Incomplete femoral line.

However, it differs from the above species by having following characters:

1. Elytral shape and colour, elytra with 2 black spots at anteriorly.
2. Mandible, immediately triangular with prominent basal tooth pointed.
3. Labrum shape.
4. Femur dark blackish brown.
5. Tegmen, median lobe shorter than lateral lobes.
6. Pronotum with 2 black spots.
7. Flagellar Formula:
   SL/W = 1.4, PL/W = 0.8, 2L/W = 1.25, 4L/W = 1.5, 7L/W = 2.3, 8L/W = 0.62, 9L/W = 1.14.
8. Sipho, siphonal capsule well developed; immediately curved siphonal tube, uniform in width.

Micraspis hypogaea sp. nov.

(Plate 27, Figs. 157-164)

Male

Body small, 6.00 mm long and 4.20 mm wide, shortly oval some what quadrate and strongly convex on dorsal side; pronotum 1.1 mm long and 3.12 mm wide; antenna 0.70 mm long; elytra brownish,

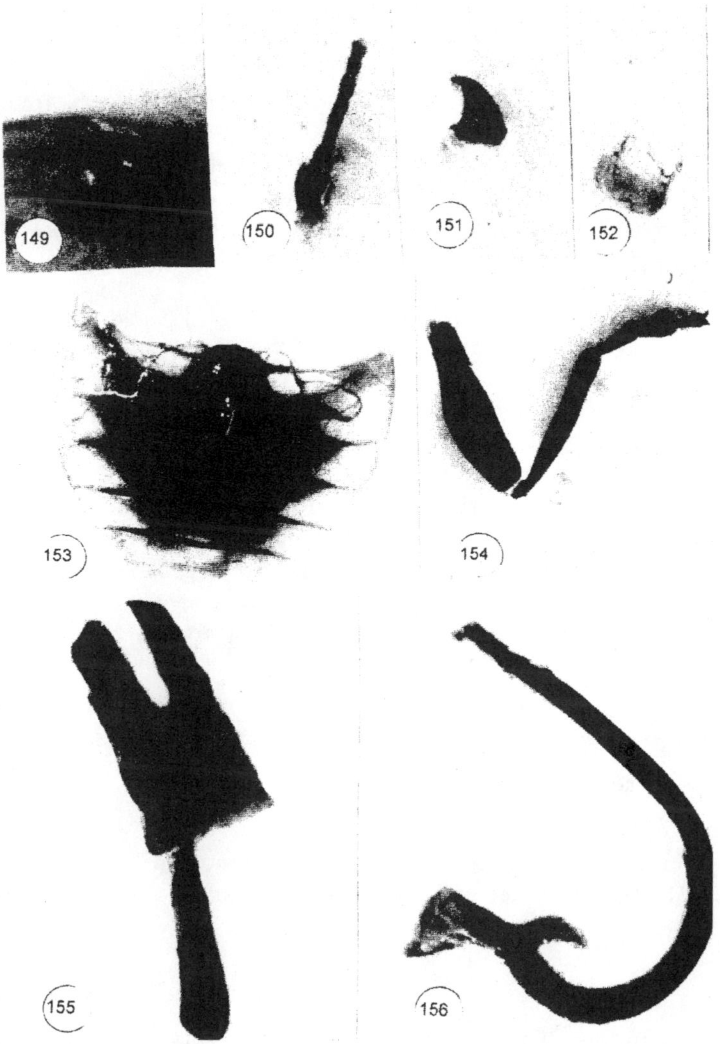

Plate XXVI: ***Micraspis gossypii*** **sp. nov. (Figs. 149-156)**

Fig. 149: Adult, Fig. 150: Antenna, Fig. 151: Mandible, Fig. 152: Labrum, Fig. 153: 1st Abdominal Sternite, Fig. 154: Hind leg, Fig. 155: Tegmen, Fig. 156: Sipho.

4.9 mm long and 4.2 mm wide, glabrous, each elytra without spots, siphonal tube 2.55 mm long.

Head

Dark brown, 1.40 mm long and 2.11 mm wide; yellowish brown, frons narrow, frontoclypeal suture not distinct; clypeus nearly straight, anterior margin and expanded infront of compound eyes; eyes black, rough and shiny, interocular distance 1.15 mm; gular sutures parallel; tentorium with parallel arms; mouth parts yellowish brown with black tinge.

Antenna (Fig. 158)

11 segmented, 0.70 mm long, short clavate, yellowish brown, including scape and last three segments; scape large, elongated, narrow basally, squarish with 8 fine setae placed on inner lateral sides, longest seta 0.14 mm; pedicel smaller than scape, more or less quadrate, longer than other segments with three setae on inner lateral sides; 1st segment longer than second; second & fourth segments more or less equal in length and width, 5th, 6th segments similar; 7th, 8th and 9th segments large; 7th segment with single seta on inner lateral side; 8th segment with 2 setae; one on each lateral side; 9th segment transverse with 9 fine long setae, last segment sparsely hairy.

Flagellar Formula

SL/W = 0.16, PL/W = 1.4, 2 L/W = 1.5, 5 L/W = 1.25, 7 L/W = 0.8, 8 L/W = 0.75, 9 L/W = 1.3.

Labrum (Fig. 161)

Somewhat elongated, laterally 0.17 mm long and 0.30 mm wide; lateral borders rounded and covered by setae; anterior side rounded, posterior side somewhat straight with two brown spots latero-basally.

Mandible (Fig. 159)

Triangular, 0.30 mm long and 0.20 mm wide at base; apical tooth bifid with 0.04 mm depression from marginal line and 0.05 mm depression from apical tooth; basal tooth small, dark brown, mandible slightly parallel sided at base; membranous postheca 0.21 mm wide.

Maxilla

Cardo 0.23 mm long and 0.25 mm wide, as wide as stipe, stipe

0.44 mm long and 0.22 mm wide; galea and lacinia 0.20 mm long with dense setae on inner side; maxillary palpus four segmented, 1.20 mm long; 1st segment small, 0.06 mm; 2nd segment 0.30 mm, narrow at base and broad at apex, longer than 3rd segment; 3rd segment 0.24 mm, more or less triangular, terminal segment 0.70 mm, large apically, truncate, more or less triangular and securiform.

Labium

Submentum fused with gula; mentum triangular, 0.19 mm long and 0.29 mm wide; prementum quadrate, 0.24 mm long and 0.39 mm wide, apically borders covered by hairs; labial palpus 3 segmented, 0.40 mm long, basal segment small, 0.02 mm; 2nd segment 0.14 mm; 3rd segment 0.19 mm and bluntly pointed.

Thorax

1.55 mm long and 3.80 mm wide; prothroax broader than long; pronotum blackish brown, 1.78 mm long and 3.12 mm wide, glabrous, anterior corners angulated with brownish spot; surface of hypomeron flat; mesosternum bisinuate, shallowly emerginate at anterior side, hind coxal cavity oval and strongly transverse; scutellum yellowish brown, triangular.

Elytra (Fig. 157)

4.9 mm long and 2.45 mm wide, brownish, glabrous, granulate with elevated margin towards outer border; elytron without spots, anterior side without spot; elytral epipleuron as wide as body width.

Hind Leg (Fig. 162)

Brownish black; coxa 0.50 mm long and 0.25 mm wide; trochanter 0.26 mm long and 0.16 mm wide, somewhat triangular with denticulate process on outer side; femur broad, 0.90 mm long and 0.34 mm wide, narrow towards apex, broader at middle, dark brown to blackish; tibia 0.90 mm long and 0.15 mm wide, cylindrical, reddish brown, basal part narrow, outer side covered by dense, thick, prominent spines; tarsi cryptotetramerous, 0.75 mm long and 0.10 mm wide; basal segment, 0.63 mm long and 0.15 mm wide, narrow at base and broad at apex with setae on outer side; 2nd segment 0.34 mm long and 0.8 mm wide, rounded apex with setae; 3rd segment very small, 0.10 mm long and 0.08 mm wide; terminal segment 0.55 mm long and 0.14 mm wide, dark brown, gradually broad towards apex and terminated into bifid claw; claw stout, 0.09 mm, curved and pointed with basal tooth.

Abdomen

3.15 mm long and 4.54 mm wide, dorsal side brownish blackish, 1.7 mm, terga slightly sclerotized; 6th sternite visible; ventral side blackish brown, sparsely punctured; femoral line incomplete, area around femoral line closely punctured.

Male Genitalia

9th tergite 'V' shaped, 0.28 mm long and 1.56 mm wide, relatively large; 10th tergite 0.19 mm long and 1.24 mm wide, strongly transverse; 9th pleurite attached to 9th tergite; 9th sternite small; apophysis elongated, 1.22 mm long and bifid at base.

Sipho (Fig.164)

2.64 mm long; siphonal capsule well developed, 0.26 mm long and 0.10 mm wide; outer lobe 0.29 mm long and 0.63 mm wide, broad; inner lobe 0.15 mm long and 0.6 mm wide, outer side slightly curved and somewhat straight at basal side; siphonal tube, 2.53 mm long and 0.5 mm wide, not uniform in width, not immediately curved but more curved with narrow apex.

Tegmen (Fig. 163)

Tegmen strut elongated broad, 0.50 mm long and 0.08 mm wide, uniform in width; lateral lobes longer than medium lobe, 0.35 mm long and 0.06 mm wide, apex densely covered by long setae; median lobe 0.30 mm long, triangular and narrow, elongated and bluntly pointed at apex.

Female

Unknown.

Holotype

Male, India, Maharashtra, Kolhapur, Coll. 4-VII-2000, Vaishali J. Patil on groundnut (*Arachis hypogaea* Linn.), antenna, labrum, mandible, maxilla, hind leg, sipho and tegmen on slides lebelled as above.

Paratype

Female 1, data same as in holotype. India, Maharashtra, Kolhapur Coll. 4-VII-2000, Vaishali J. Patil.

Prey

Aphid, *Aphis craccivora* Koch; *A. citricola*.

Prey Plant

Groundnut, *A. hypogaea*, Citrus, *Citrus* sp.

Discussion

This species runs close to *Micraspis discolor* (Fabricius) by having following characters:

1. Maxilla.
2. Labium.
3. Complete straight coxal line.
4. Incomplete femoral line.
5. General antennal shape.
6. Hind legs similar.
7. Eyes same.

However, it differs from the above species by having following characters:

1. Eyltral shape and colour, without spots.
2. Mandible parallel sided at base, strongly bifid and accurately pointed.
3. Maxillary palpus segments.
4. Femur dark blackish brown.
5. Siphonal tube not uniform in width, siphonal capsule well developed.
6. Labrum shape and spots.
7. Flagellar Formula:
   SL/W = 0.16, PL/W = 1.4, 2L/W = 1.5, 5L/W = 1.25, 7L/W = 0.8, 8L/W = 0.75, 9L/W = 1.3.

GENUS *VERNIA* MULSANT, 1850

Coccinella discolor Fabricius, 1798, *Suppl. Ent. Syst.*, 77 *Vernia discolor* Mulsant, 1850, *Spec. Irim. Securipalp:* 369.

Type Species: *Vernia discolor.*

The genus is erected by Mulsant in 1850 and it includes 4 species *viz.; V. discolor, V. kiotoensis, V. cardoni and V. polyphagi* (Sathe & Bhosale, 2001). Omkar and Bind (1992) reported *V. cardoni* from

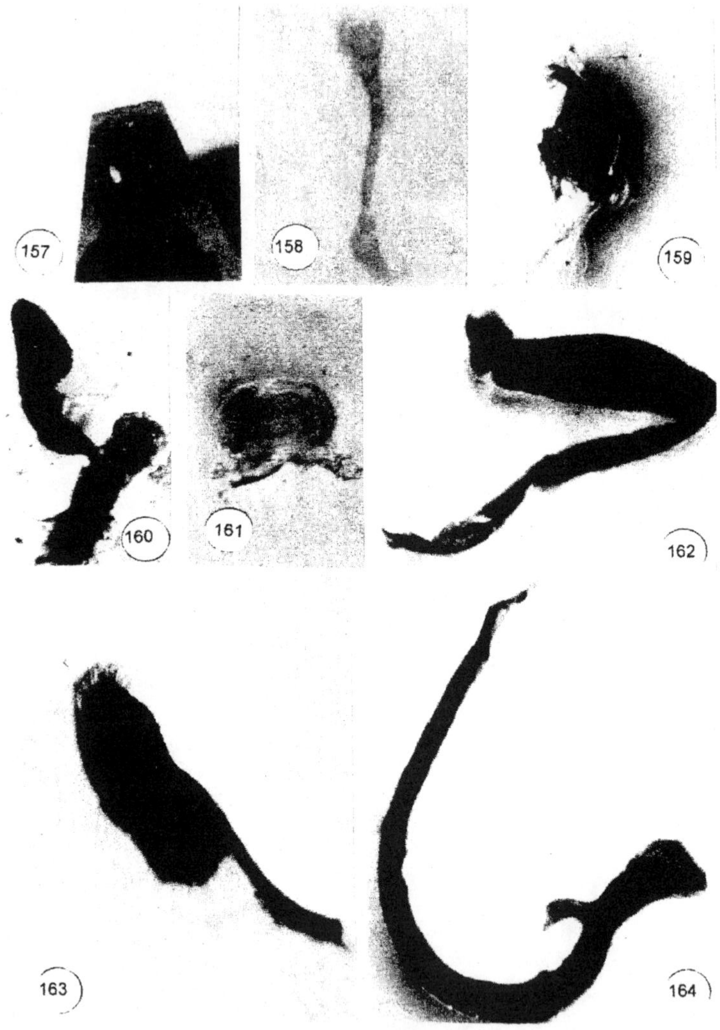

Plate XXVII: *Micraspis hypogaea* sp. nov. (Figs. 157-164)

Fig. 157: Adult, Fig. 158: Antenna, Fig. 159: Mandible, Fig. 160: Maxilla, Fig. 161: Labrum, Fig. 162: Hind leg, Fig. 163: Tegmen, Fig. 164: Sipho.

Uttar Pradesh. The genus is represented by body hermispherical, glabrous; dorsum orange brown; frons sometimes with black spot; discal spots on pronotum; black scutellum; elytron with very narrow black sutural area; femora brown to black; tibia and tarsi brown; pro-sternal carinae extending half of prosternal length. From India recently, Sathe & Bhosale (2001) described a new species, *V. polyphagi* feeding on aphids, *R. maidis.*

Vernia vulgarae sp. nov.

(Plate 28, Figs. 165 to 172)

Female

5.3 mm long and 4.80 mm wide, oval, dorsum convex, underside black; pronotum 1.05 mm long and 2.93 mm wide; antenna 0.55 mm long, 11 segmented; elytra yellowish brown, 4.75 mm long and 2.25 mm wide; spermathica yellowish "V" shaped, 0.42 mm long and 0.08 mm broad, narrow at middle, broader at apex.

Head

1.54 mm long and 1.79 mm wide, quadrate, inserted into pronotum, with dark brown apices of mandibles; frons with black spot; fron to clypeal suture indistinct; clypeus broad expanded infront of eyes; eyes laterally projected, interocular distance 0.9 mm.

Antenna (Fig. 166)

11 segmented 0.55 mm long, scape large, quadrate with setae, tapering slightly distally, slightly produced inwardly; pedicel elongated narrower anteriorly with setae; 1st to 5th segments more or less similar but first narrower basally and second more or less parallel sided and each with single seta; 6th also more or less similar to 5th segment; last 7th, 8th and 9th segments larger than 1st to 6th segments and serrated, 9th segment elongated, more or less oval, tapering towards both the ends, with numerous setae.

Flagellar Formula

SL/W = 0.87, PL/W = 1, 2L/W = 2, 4L/W = 1.6, 7L/W = 1.3, 8L/W = 12, 9L/W = 1.3.

Labrum (Fig. 169)

0.15 mm long an 0.25 mm wide, somewhat quadrate with; Lateral borders rounded, anterior border covered by setae, yellowish coloured with dark spot.

Mandible (Fig. 167)

0.25 mm long, 0.19 mm wide; bifid apical, tooth, triangular, bifid apical tooth dark reddish brown, pointed and curved; depression from apical tooth 0.05 mm, 0.06 mm depression from marginal line, 0.19 mm membraneous prostheca; basal tooth single, pointed, yellowish, prostherca yellowish brown; mandible not tapering immediately from the base; basally more or less parallel sided for some distarlce.

Maxilla (Fig. 168)

Cardo as long as wide, 0.3 mm long and 0.28 mm wide; stipe longer than wide; galea 0.27 mm long spatulate, covered by setae; lacinia longer than galea, elongated, simple, inner side covered with fine setae; 4 segmented maxillary palpus; 0.09 mm long, 1st segment very small, 2nd not smaller than 3rd segment, 0.81 mm long; 3rd segment, 0.29 mm long and 0.27 mm wide, large terminal segment, apically truncate and securiform, not elongately 'triangular'.

Labium

Submentum fused with gala; mentum quadrate, small; 0.16 mm long and wide; apically broad and trapezoidal prementum, 0.15 mm long and 0.23 mm wide; 3 segmented labial palpus; very small 1st segment, 0.03 mm; 2nd segment tapering, rounded at end, 0.16 mm long.

Thorax

1.13 mm long and 3.1 mm wide; pronotum yellowish brown, 1.3 mm long and 2.98 mm wide, rounded lateral sides, densely punctured, basal border with black marking joins with anterior black marking by vertical black strip; half posternal carinae; mesosternum broader than long; metandosternite with pair of fulcral arms, scutellum dark brown.

Elytra (Fig. 165)

4.75 mm long and 2.50 mm wide, yellowish brown, glabrous, distinctly sparsely punctured; inner border black, elytra with two and half black spots, small spot anteriorly, another larger black spot towards mid-outer side, spot elongated laterally, half black spot at the anterior side; basal border slightly wider than pronotum outer margin included more towards apex.

Hind Leg (Fig. 172)

Coxa 0.47 mm long and 0.18 mm wide, stout, strong, dark reddish brown at posterior side, somewhat rounded; trochanter, 0.22 mm long and 0.16 mm wide; somewhat triangular, small with denticulate process; femur 0.70 mm long and 0.30 mm wide, elongated. Swollen at anterior side, anterior side convex, inner side somewhat straight, yellowish brown; tibia yellowish more elongated than femur, 0.76 mm long and 0.12 mm wide with tibial spines and setae, tibial spurs not prominent; tarsi 0.35 mm long and 0.8 mm wide, 3 segmented; yellowish brown; 1st segment laterally covered by setae, broad towards apex, triangular; 2nd segment laterally covered by setae, elongated; elongated terminal segment, broad towards apex, terminate into bifid claw; Claw 0.30 mm, basal tooth pointed, curved.

Abdomen

2.36 mm long and 3.2 mm wide, dorsum black with yellow border; 6th sternite visible; lateral side covered by setae; hairy ventrum; complete coxal line on 1st abdominal sternite (Fig. 171); complete and slightly angulate femoral line, hairy area around femoral line.

Female Genitalia

Hemistemite elongated with stylus. Spermathica (Fig. 170), yellowish, "V" shaped, 0.40 mm long and 0.8 mm broad, curved with weak marking on surface, narrower at middle, rounded at ends, broader at apex.

Male

Unknown.

Holotype

Female, India, Maharashtra, Hatkanangale, Coll. 16-VIII- 2001, Vaishali J. Patil, on Hybrid jowar, antenna, labrum, mandible, maxilla, hind leg, spermathica on slides labelled as above.

Paratype

1 Female, Coll. T. V. Sathe, 10-VIII-2000, India: Maharashtra, Kolhapur, data same as holotype.

Prey

Aphid, *Rhophalosiphum maidis* (Fitch).

Prey Plant

Hybrid Jowar (*Sorghum vulgare* Linn.).

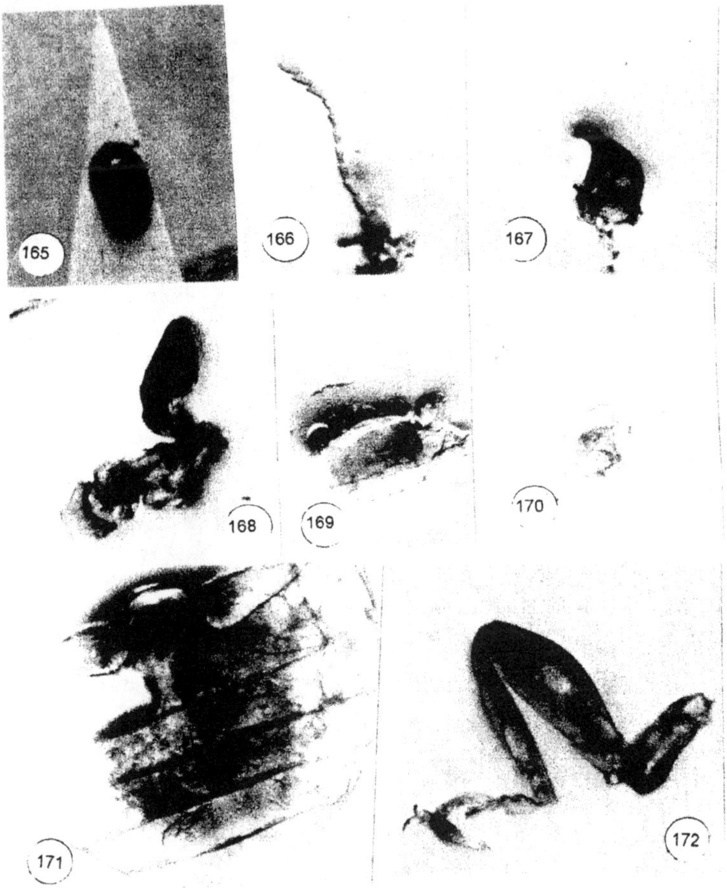

Plate XXVIII: *Vernia vulgarae* sp. nov. (Figs. 165-172)

Fig. 165: Adult, Fig. 166: Antenna, Fig. 167: Mandible, Fig. 168: Maxilla, Fig. 169: Labrum, Fig. 170: Spermathica, Fig. 171: 1st Abdominal Sternite, Fig. 172: Hind leg.

Discussion

Vernia vulgarae sp. nov. runs close to *V. polyphagi* (Sathe & Bhosale, 2001) by following characters:

1. Antennal colouration.
2. Mandible bifid apically.
3. Abdominal 6th tergite hairy.

However, it differs from above species by following characters:

1. Antennal tip larger and serrated, first flagellar segment narrower basally round, segment more or less parallel sided.
2. Flagellar Formula:
   SL/W = 0.87, PL/W = 1, 2L/W = 2, 4L/W = 1.6, 7L/W = 1.3, 8L/W = 12, 9L/W = 1.3.
3. Segments of maxillary palpus.
4. Mandible parallel sided basally and then tapers.
5. Spermathica 'V' shaped.
6. Typical elytral spots two and half.

Taxonomy of Carabid Beetles

Family: Carabidae

The family Carabidae is one of the largest family in the order Coleoptera of the Class Insecta. Nearly, 21,000 species are known from the world and many more are to be described. Commonly, they are found on the grounds, their elytra become fixed and hind-wings are atrophied. They are also plenty in tropics where they are ground-dewellers and also arboreal in tropics. Hind-wings and elytra of these beetles become free. These beetles are commonly known as ground beetles. The members of this great family are mostly somber in appearance and a few are brightly coloured. Majority of them are nocturnal in habit and active during night-time. They usually live under bark, stone, logs, rotten wood specially on damp places, their larvae and adult feed on other insects. In India, they found in all places from top mountain area to sea level. The approximate number of these beetles in India appear to be 2,000 species and sizes vary from 1.5 mm to 22 mm.

Andrews (1930) classified the family Carabidae into two subfamilies Carabiniae and harpalinae and sub families were divided into following tribes, Ozaenini, Scaritini, Siagonini, Elaphrini, Loricerini, Omophronini, Opisthiini, Nebriini, Notophilini, Carabini, Hiletini, Hexagoniini, Granigerini, Pterostichini, Bembidiini, Dispheaericini, Broscini, Idiomorphini, Harpalini, Trechini, Anchomenini, Odacanthini, Brachinini, Apotomini, Licinini, Panagaeini, Chlaeniini, Tetragonoderini, Lebiini, Pentagonicini, Helluonini, Physocrotaphini, Orthogniini, Zuphiini, Dryptini, Anthiini and Pseudomorphini. Ball (1979), Lindorth (1969) attempted to classify Carabidae into more sub-families by giving status of sub-families to different tribes of Andrews (1930, 1935). The same classification is followed in the present study. In past Horn (1915), Sloane (1923), Andrews (1929, 1930, 1935), Moore (1963), Pearson (1988) and Saha et at., (1992) worked on taxonomy of carabids.

Tribe - Anthiini

Genus *Anthia* Horn

The genus *Anthia* is represented by only two species from India *viz. A. sexgutatta* and *A. carabus* (Andrews, 1930, 1935). The species described under the genus *Anthia* is newly added to this genus.

The genus *Anthia* is characterized by following characters: Body often black/brown, sometimes with white paths on elytra and thorax, longer than wide; elytra fitting tightly to the body; hard thick, hairy, oval, more flattened, compound eyes large; antenna fifiform, mandible strong, maxilla not hooked, clypeus infront of antennae; tarsal joints distinct; claw well formed.

Anthia kolhapuriensis sp. nov.

(Plate 29 Figs. 173 to 180)

Male

Body oval, 45 mm long and 15 mm wide, large black; antenna 20 mm long: elytra 25 mm long and 15 mm wide, blackish with 4 white spots; sipho 11 mm. long.

Head

13 mm long and 9 mm wide, large, blackish, frons blackish, clypeus not extending laterally infront of base of antennae; eyes large, blackish.

Antenna (Fig. 175)

11 segmented, 20 mm long, filiform, not elbowed and projecting conspicuously in front of head; scape large, longer than wide, whitish black; pedicel smaller than scape, 1st segment longer than other segments; 2nd, 3rd, 4th and 5th segments more or less similar in shape and size, 6th, 7th and 8th also similar somewhat rectangular, terminal segment smaller than other segments, narrow towards apex.

Flagellar Formula

SL/W = 4, PL/W =- 2, 2L/W = 2.37, 4L/W = 2.27, 7L/W = 2.25, 8L/W = 2.16, 9L/3 = 2.94.

Labrum (Fig. 176)

5 mm long and 2 mm wide, elongated, oval, blackish.

Mandible (Fig. 177)

10 mm long and 4 mm wide, curved, pointed, elongated triangular, inner side somewhat straight; basal tooth present, not pointed, blackish; membranous prostheca 3 mm long.

Maxilla (Fig. 174)

Blackish, inner side hairy, maxillary palpus 4 segmented, 1st segment shorter, 2nd segment longer than other segments, elongated, narrow at base and broad at apex; 3rd segment elongated, small, narrow at base and broad at apex; 4th segment apex rounded, broad, elongated.

Thorax

11 mm long and 15 mm wide, prothorax large, blackish with 2 white elongated spots antero-lateral side; prothorax hairy; thorax wider than head, anteriorly wider than posterior. "Y" shaped.

Elytra (Fig. 173)

25 mm long and 15 mm wide; blackish with four more or less rounded white spots, anterior larger than posterior; elytra fitting tightly to the body, rows of pits absent on elytra; oval, rounded anteriorly and posteriorly; elytra thick and hard, hairy.

Hind Leg (Fig. 178)

Coxa 4 mm long and 3 mm wide, blackish, hairy, broad at base and narrow at apex; trochancter 4.5 mm long and 2.2 mm wide, elongated, oval, flat, blackish, hairy; femur 13 mm long and 3.5 mm wide, elongated, outer side somewhat straight, hairy, inner side broad, at middle, blackish, densely hairy, broad at apex and narrow at base; tibia 14 mm long and 0.90 mm wide, longer, blackish, small densely hairy, tibia shows one long spine; tarsi five segmented, 4.7 mm long, 1st segment longer than other segments, elongated, covered with small dense hairs; 2nd, 3rd and 4th segments more or less similar also covered with hairs; terminal narrow at base and broad at apex, terminated in bifid, curved, pointed, claw; claw 0.98 mm long, curved, blackish, pointed.

Abdomen

21 mm long and 12 mm wide, flat, blackish, hairy, coxal line straight, 6th sternite visible; last segment of abdomen rounded, black (Fig. 179).

Male Genitalia

9th tergite rounded, posteriorly, blackish.

Sipho (Fig. 180)

11 mm long, broad, large, blackish brown, apex narrow, siphonal capsule 4.3 mm long and 3 mm wide; siphonal tube broad, 8 mm long and 3 mm wide, slightly curved, blackish, narrow at apex.

Female

Unknown.

Holotype

Male, India: Maharashtra, Kolhapur, Coll. 20-VIII-2002, Vaishali J. Patil, on grass land, antenna, Labrum, mandible, maxilla, hind leg and sipho of male on slides labelled as above.

Prey

Grasshoppers (*Hieroglyphus banian* Fab.), sugarcane wooly aphids, small caterpillars, cutworms.

Prey Ecosystem

Grassland ecosystem.

Discussion

This species runs close to *Anthia sexgutatta* Horn by having following characters, antenna, mandible, clypeus, eyes, maxilla, legs. However, it differs from the above species by the following characters:

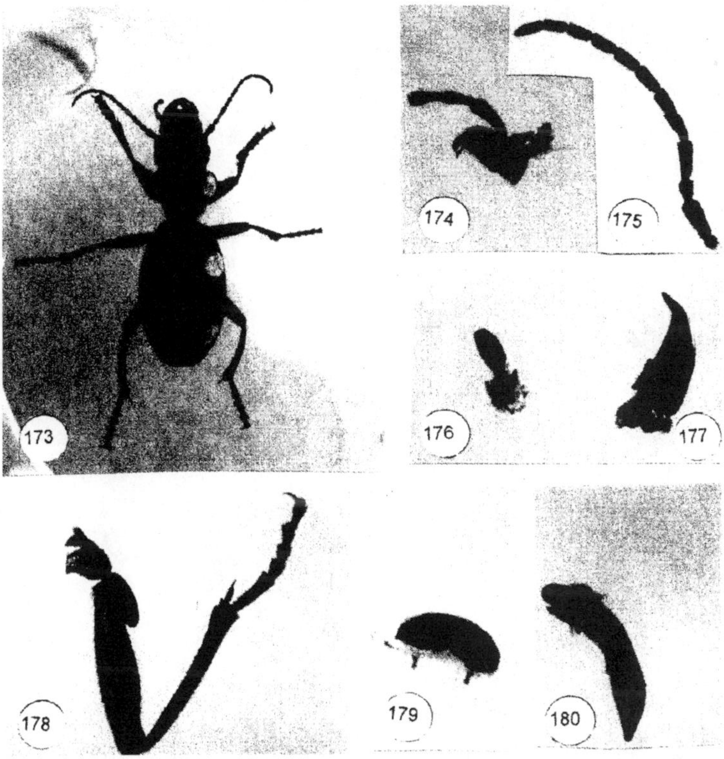

Plate XXIX: *Anthia kolhapuriensis* sp. nov. (Figs. 173-180)

Fig. 173: Adult, Fig. 174: Maxilla, Fig. 175: Antenna, Fig. 176: Labrum, Fig. 177: Mandible, Fig. 178: Hind leg, Fig. 179: Last Abdominal Segment, Fig. 180: Sipho.

1. Elytral spots white rounded and four in number instead of 6 in *A. sexgutatta*.
2. On prothorax rounded two white spots instead of elongated in *A. sexgutatta.*
3. Labrum.
4. Sipho.
5. Flagellar Formula:

   SL/W = 4, PL/W -2, 2L/W = 2.37, 4L/W = 2.27, 7L/W = 2.25, 8L/W = 2.16, 9L/W = 2.94.

GENUS *EUDEMA* HORN. (1915)

Eudema angulatum Horn.

(Plate 30, Figs. 181 to 185)

Male

Body oval, 15 mm long and 7 mm wide, blackish; antenna 9 mm long; elytra, 10 mm long and 3.5 mm wide, elytra black with four irregular yellow patches; sipho 1.98 mm long.

Head

4.6 mm long and 2.5 mm wide, blackish, large; frons blackish, clypeus not extending laterally infront of base of antennae; eyes large, blackish.

Antenna (Fig. 182)

11 segmented, 9 mm long, filiform, not elbowed and projecting conspicuously in front of head; scape larger than pedicel; with fine setae at anterior side; pedicel small somewhat rectangular covered with fine setae; 1st segment longer than other segments, elongated, blackish, covered by long fine setae, 2nd segment shorter than 1st segment, broad, covered with short fine setae; 3rd to 8th segment more or less rectangular, dark blackish, covered with small hairs, narrower at the end; terminal segment elongated, narrow at apex.

Flagellar Formula

SL/W = 2.2, PL/W = 1.7, 2L/W = 2, 4L/W = 2.2, 7L/W = 2.9, 8L/W = 2.5, 9L/W = 4.2.

Labrum

0.50 mm long and 0.90 mm wide, reddish, brown, wider than long.

Mandible (Fig. 185)

0.78 mm long and 0.32 mm wide, pointed apical tooth, apical tooth short reddish black; depression 0.13 mm from marginal line; membranous prostheca 0.35 mm, dark reddish blackish; basal tooth pointed, reddish brown, hairs absent on basal tooth.

Maxilla

Reddish brown to black, outer side hairy, maxillary palpus 4 segmented (Fig. 183); 1st segment very short, dark blackish brown; 2nd segment longer than other segments, elongated, narrow at base and apex, anterior portion covered with long fine setae, dark brownish black' 3rd segment shorter than 1st segment, broad at apex and narrow at base, covered with long setae, brownish black; last terminal segment broad, more or less triangular, laterally covered with small setae, broad at apex and narrow at base.

Thorax

4 mm long and 5 mm wide, dark blackish, prothorax large, without spot; short hairs present on thorax; thorax wider and longer than head; posterior corners rounded, posteriorly wider than anterior.

Elytra (Fig. 181)

10 mm long and 3.5 mm wide; blackish, having four irregular yellow patches on elytra; elytra fitting tightly to the body with rows pits/lines, elytra elongated oval, covered by fine hairs; anteriorly yellow spots broader than posteriorly.

Hind Leg

Coxa 1.7 mm long and 0.90 mm wide, blackish, rounded; trochanter 1.8 mm long and 1.2 mm wide, longer than wide, elongately rounded; femur 4 mm long and 0.50 mm wide, blackish, covered with small hairs, elongated, more or less straight borders; tibia 4.5 mm long and 0.20 mm wide, elongated, tibia shows long spine at anterior base; tarsi five segmented, 3.10 mm long and 0.08 mm wide; 1st segment longer than other segments, elongated, covered with small hairs; 2nd, 3rd and 4th segments more or less similar also covered with hairs; terminal segment blackish, termines into bifid claw; pointed claw, claw 0.12 mm long.

Abdomen

5.4 mm long and 3.6 mm wide, flat, blackish, hairy, coxal line straight, 4 visible segments.

Male Genitalia

9th tergite narrow.

Sipho (Fig. 184)

1.98 mm long and 0.70 mm wide, siphonal tube short, elongated cured at tip and pointed.

Holotype

Male, India: Maharashtra, Kolhapur. Coll. 15-VII-2001, T.V. Sathe, in grassland, antenna, mandible, hind leg and sipho of male on slides labelled as above.

Paratype

Male 2, data same as in holotype. India: Maharashtra, Kolhapur, Coll. 20-VIII-2002. Vaishali J. Patil.

Preys

Smaller caterpillars, cutworms, wire worms; jassids, grasshoppers, crickets, silver fishes, aphids, wolly aphids, cockroaches, bed bug, etc.

Prey Ecosystem

Grassland ecosystem and open ecosystem.

Eudema polyphagi sp. nov.

(Plate 31, Figs. 186 to 192)

Male

Body oval elongately, 28 mm long and 13 mm wide, blackish brown, covered with fine hairs; antenna 12 mm long; elytra 18 mm long 0.6 mm wide, elytra blackish brown with four large irregular dark yellow patches; sipho 2.10 mm long.

Head

7.6 mm long and 3 mm wide, brownish black, covered with fine hairs, broader at posterior, large; frons blackish, clypeus not

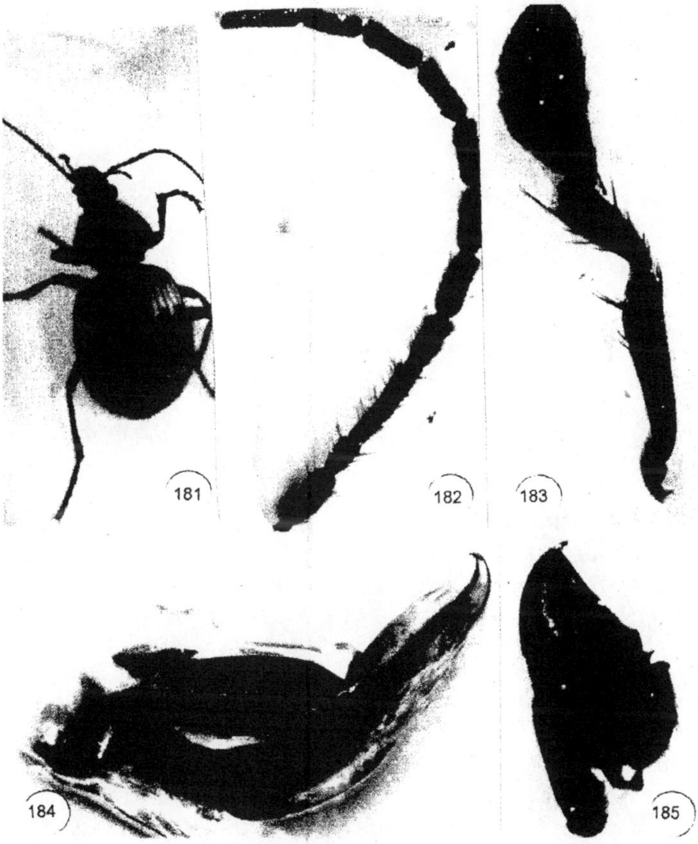

Plate XXX: ***Eudema angulatum*** **(Horn) (Figs. 181-185)**

Fig. 181: Adult, Fig. 182: Antenna, Fig. 183: Maxillary Palpi, Fig. 184: Sipho, Fig. 185: Mandible.

extending laterally infront of base of antennae; eyes prominent, large, blackish.

Antenna (Fig. 187)

11 segmented, 12 mm long, filform, not elbowed and projecting in front of head; scape longer than pedical, elongated, blackish, covered with fine hairs; pedicel shorter than scape, covered with fine hairs; flagellar 1st segment longer than other segments, elongated; 3rd to 8th segments more or less similar, blackish.

Flagellar Formula

SL/W = 2.7, PL/W = 2.25, 2L/W = 2.8, 4L/W = 3, 7L/W = 2.9, 8L/W = 2.7.

Labrum (Fig. 190)

1.3 mm long and 0.45 mm wide, brownish black, wider than long, posterior side straight and anterior side curved, lateral side rounded curved.

Mandible (Fig. 188)

0.90 mm long and 0.55 mm wide, triangular; apical tooth pointed, not bifid, apical tooth longer, acute apex, depression 0.10 mm from marginal line; two basal tooth present brownish black, covered with fine hairs; membranous prostheca 0.42 mm; dark brownish black.

Maxilla (Fig. 189)

1.3 mm long and 0.35 mm wide, brownish black, outer side covered with hairs; maxillary palpus 4 segmented, 1st segment short, brownish; 2nd segment longer than other segments, blackish, covered with long fine hairs, elongated, broadly curved at anteriorly, narrow at base; 3rd segment shorter than 2nd segment, narrow at base and broad at apex, with fine hairs; 4th or terminal segments triangular, narrow at apex and base, covered with small fine hairs, broad.

Thorax

8.8 mm long and 10 mm wide, brownish black, prothorax, large, somewhat triangular, posteriorly broad, more or less rounded lateral borders; lateral borders covered with fine hairs.

Elytra (Fig. 186)

18 mm long and 0.6 mm wide, brownish black with four large

irregular dark yellow patches; elytra fitting tightly to the body with rows/pits, elongated oval, somewhat convex, covered with fine hairs; anterior spots large and dark yellow reaching at lateral edges; posterior spots small, yellowish, not touching to lateral sides.

Hind Leg (Fig. 191)

Coxa 1.8 mm long and 1.2 mm wide, blackish, rounded, small; trochanter 2 mm long and 1.5 mm wide, longer than wide, rounded elongately, femur 7 mm long and 1 mm wide, blackish brown, with fine hairs, elongated, straight lateral borders; tibia 8 mm long and 0.90 mm wide, elongated with spine at anterior base; tarsi five segmented, 5 mm long and 0.07 mm wide, 1st segment longer than other segments, elongated with hairs; 2nd, 3rd and 4th segments shorter than 1st segment, brownish black with fine hairs; terminal segment terminated into bifid pointed claw; claw 0.14 mm long, pointed, curved.

Abdomen

11.6 mm long and 4.7 mm wide, blackish, hairy, coxal line straight, 4 segments visible.

Male Genitalia

9th tergite narrow.

Sipho (Fig. 192)

2.10 mm long; siphonal tube dark black, long with blunt apex; inner lobe dark reddish brown.

Female

Unknown.

Holotype

Male, India: Maharashtra, Kolhapur. Coll. 7-IX-2000, Vaishali J. Patil, in grass land ecosystem, antenna, mandible, maxilla, labrum, hind leg and sipho of male on slides labelled as above.

Paratype

Male 1, data same as in holotype, India: Maharashtra, Kolhapur, Coll. 18-IX-2001, T.V. Sathe.

Prey

Grasshopers, cutworms, jassids, aphids, crickets, small caterpillars etc.

Prey Ecosystem

Grassland ecosystem and open ecosystem.

Discussion

This species runs close to *E. angulatum* (Horn) by having following characters, antenna, maxilla, labrum, hind leg. However, it differs from above species by having following characters:

1. Mandible not with bifid apical tooth, with acute apex, two basal tooth present, brownish black, covered with hairs.

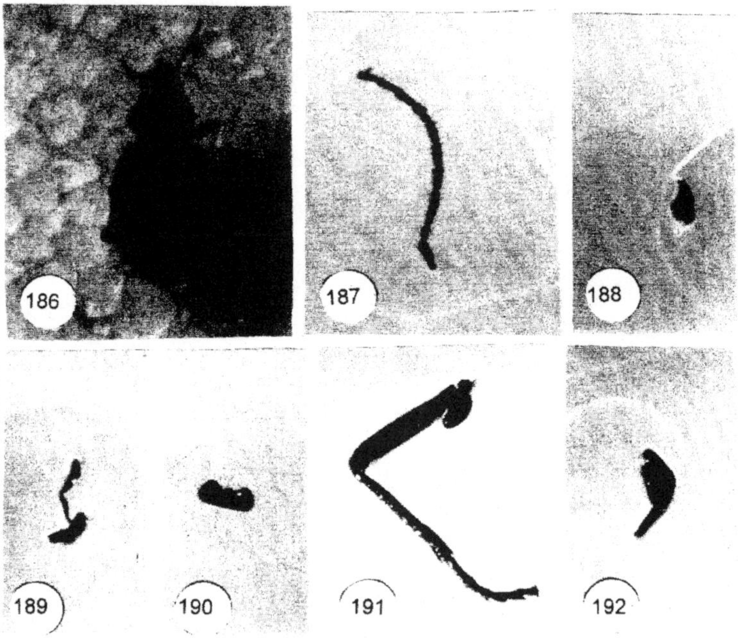

Plate XXXI: *Eudema polyphagi* sp. nov. (Figs. 186-192)

Fig. 186: Adult, Fig. 187: Antenna, Fig. 188: Mandible, Fig. 189: Maxilla, Fig. 190: Labrum, Fig. 191: Hind leg, Fig. 192: Sipho.

2. Body size smaller than *E. angulatum*.
3. Elytral spots.
4. Sipho, siphonal tube long with blunt apex.
5. Flagellar Formula:

   SL/W = 2.7, PL/W = 2.25, 2L/W = 2.8, 4L/W = 3, 6 L/W = 2.9, 8L/W = 2.7.

GENUS *DICRANONCUS* BATES, 1892

Ann. Mus. Star. nat. Genova (1892)

Dicranoncus indicus sp. nov.

(Plate 32, Figs. 193 to 198)

Female

Body somewhat rectangular, flat, yellowish reddish brown, 10 mm long and 5 mm wide; antenna 7 mm long; elytra 7 mm long and 2.5 mm wide; blackish brown with two yellowish spots; sepermathica 0.35 mm long.

Head

1.5 mm long and 1 mm wide, reddish brown, somewhat rounded, Head small, convex; frontal impressions small, shallow; eyes rather large, but flat, forming with genae, continuous curve to the neck, lightly constricted as sides black.

Antenna

11 segmented, 7 mm long, antennae long and slender, but not reaching to middle of elytra, projecting in front of head; scape short, somewhat rounded, covered with hairs; 1st segment longer, straight elongated; 2nd segment more or less straight and shorter than 1st segment; 3rd, 4th and 5th segments dark brown, slightly curved at basally; covered with fine hairs; 6th, 7th and 8th more or less similar in length, dark brown, elongated; 9th segment elongated, straight, narrow at apex, dark brown.

Flagellar Formula

SL/W = 2.35, PL/W = 4.6, 2L/W = 2.3, 4L/W = 2.35, 6L/W = 3.2, 8L/W = 3, 9L/W = 4.2.

Mandible (Fig. 194)

0.58 mm long and 0.27 mm wide, more elongated, triangular reddish brown; apical tooth blunt pointed and immediately curved, depression 0.10 mm from marginal line; basal tooth multidenticulate, dark blackish brown, below basal tooth small hairs present; membranous prostheca 0.20 mm, dark brown.

Maxilla (Fig. 195)

0.60 mm long and 0.15 mm wide, yellowish brown; maxillary palpus 4 segmented; 1st segment short, brown; 2nd segment longer than 1st segment, elongated, broader at middle, dark brown, covered with four setae laterally; 3rd segment narrow at base and broad at apical, dark brown, covered with few setae; terminal segment somewhat cylindrical, flat at anterior border, covered with fine setae, dark brown.

Labium

0.70 mm long and 0.30 mm wide, yellowish, dark brown; labial palpus 3 segmented; 1st segment short, yellowish brown; 2nd segment elongated cylindrical, narrow at end, brown, covered with fine setae; terminal segment narrow at posterior end and broad at anterior end, dark brown.

Thorax

2.5 mm long and 1.5 mm broad, reddish brown, flat, anteriorly broad and posteriorly small, anterior lateral side shows acute curved; prothorax rather flat, wider than head and longer than wide, basal sides oblique, bisetose, widely explanate and moderately reflexed, rounded in front, nearly straight behind, hind angles reflexed and strongly rounded; median line fine.

Elytra (Fig. 193)

7 mm long and 2.5 mm wide, blackish brown with four yellow spots, elongate, oval, wider and longer than thorax; base emarginate, the border sharply angulate at shoulder, slightly sinuate on each side near and minutely mucronate at apex; striae fairly deep, impunctate, elytra shorter than abdomen; elytral anterior spot larger, laterally placed; posterior spot somewhat rounded and centrally placed.

Hind Leg (Fig. 197)

Coxa 0.30 mm long and 0.25 mm wide, sub- rounded, brownish, small; trochanter 0.35 mm long and 0.28 mm wide, elongately rounded, yellowish brown; femur 1.8 mm, dark brown, hairy, elongated, narrow at apex; tibia 1.7 mm long and 0.18 mm wide yellowish brown, elongated, somewhat straight, dark brown at anterior end; tarsi 1.45 mm long, 5 segmented, 1st segment elongated, brown longer than other segment; terminal segment elongated dark brown, terminated into bifid claw; claw 0.12 mm long, curved, dark blackish brown with pointed apex.

Fore Leg (Fig. 196)

Special mark: Clining apparatus on middle tibia.

Abdomen

5.5 mm long and 4.6 mm wide; blackish brown, pointed at base; coxal line straight.

Female Genitalia

9th tergite pointed.

Spermathica (Fig. 198)

0.35 mm long and 0.20 mm wide; yellowish, narrow at one end and broad at another end.

Male

Unknown.

Holotype

Female, India: Maharashtra, Kolhapur, Coll. 4-IX-2002, Vaishali J. Patil, in grassland, open ecosystem, antenna, mandible, maxilla, labium, hind leg, and spermathica of female on slides labelled as above.

Paratype

Female, 1 data same as in holotype, India: Maharashtra, Kolhapur, Coll. 10-IX-2001, Vaishali J. Patil.

Prey

Cutworms, small caterpillars, jassids, aphids, grasshoppers etc.

Prey Ecosystem

Grassland and open ecosystem.

Discussion

This species runs close to *Dicranoncus* (*Monacanthonyx*) *pocillator* Horn by having similar characters in head and prothorax.

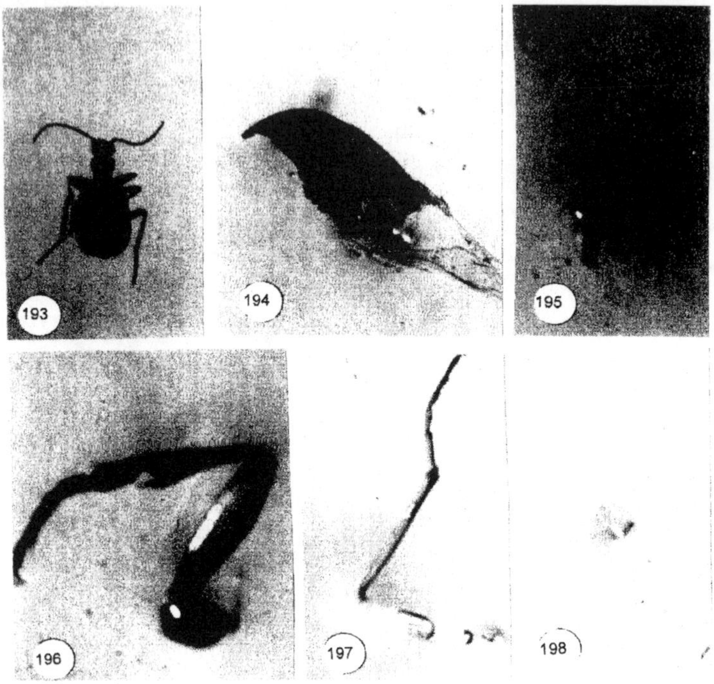

Plate XXIII: *Dicranoncus indicus* sp. nov. (Figs. 193-198)

Fig. 193: Adult, Fig. 194: Mandible, Fig. 195: Maxilla, Fig. 196: Fore leg, Fig. 197: Hind leg, Fig. 198: Spermathica.

However, it differs from the above species by having following characters:

1. Elytra blackish brown with four yellow spots.
2. Mandible triangular, reddish brown, apical tooth bluntly pointed and immediately curved; basal tooth multi denticulate.
3. Maxilla yellowish brown.
4. Spermathica yellowish narrow at one end and broad at another end.

5. Flagellar Formula:
   SL/W = 2.35, PL/W = 4.6, 2L/W = 2.3, 4L/W = 2.35, 6L/W = 3.2, 8L/W = 3, 9L/W = 4.2.

PRAYING MANTIDS

Mantids frequently live in grassland and forests of India (Mukherjee & Hazra, 1993). Praying mantids are carnivorus insects showing peculiar habits of prey-capture, camouflage and reproduction. They are well known as agent of biological control of pest insects. These insects are found in warm moist tropical rainforest, particularly after monsoon in Eastern India. These insects generally prefer altitude of about 3500 to 4000 ft. As mantids are mostly weak fliers, they are generally found in herbs and shrubs. But bigger forms are usually found on trees. They are mostly found during early part of the day on leaves or open areas of plant where other insects often visit. They are also attracted to light. During day the mantids can be collected by hand-picking or by hand net and by other general methods in use. Cannibalism is very frequent among them, particularly during a successful mating. Territoriality also exhibited by exposure of marks on body or by making sound as early as the oligocene period.

A natural suprageneric classification of mantid is difficult due to lack of adequate morphological characters. Mantids were previously considered to belong to order Dictyoptera but, Burmeister (1838) considered them under order Mantoidea for their distinctive features. They are close to Isoptera. Giglo-Tos (1927) started from family status for mantids and divided it into thirty two sub-families. However, most authors now accepted Beier's (1964) classification which is followed here. According to his classification Blattopteriodea is super order and Mantoidea is the order of mantids. A total number of 160 species under 67 genera are so far known from India (Mukherjee, 1990). Recently, Mukherjee & Hazra (1993) added 23 species to this order from India. There are 6 families of Mantids known from India out of which 3 families occur in Maharashtra, are represented in the text.

Family - Hymenopodidae

External spines of anterior tibia numerous and very close beet, forewing with eye like spiral marks.

Plate XXIII

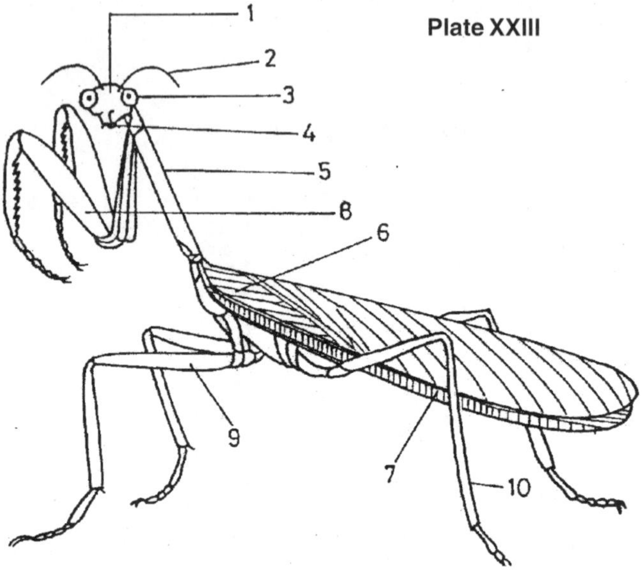

Fig. 199: Adult in Characteristic Attitude

(1) Head, (2) Antenna, (3) Eye, (4) Mouthparts, (5) Thorax, (6) Wings, (7) Abdomen, (8) Fore legs, (9) Middle legs, (10) Hind legs.

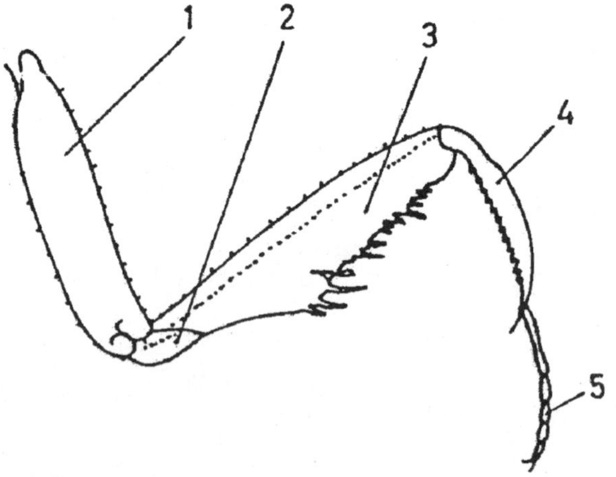

Fig. 200: Raptorial Fore Leg

(1) Coxa, (2) Trochanter, (3) Femur, (4) Tibia, (5) Tarsus.

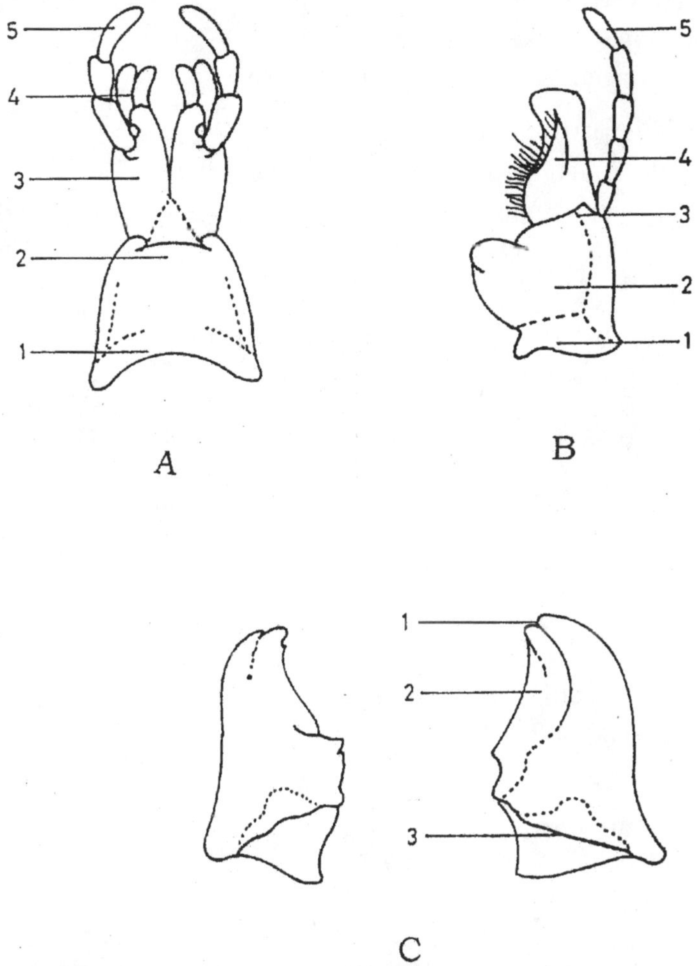

Plate XXXIV

Fig. 201: Mouth Parts

(A) *Labium*: (1) Mentum, (2) Submentum, (3) Paraglossa, (4) Glossa, (5) Labial pulp.

(B) *Maxilla*: (1) Cardo, (2) Stipes, (3) Lacinia, (4) Galea, (5) Maxillary palp.

(C) *Mandible*: (1) Apical tooth, (2) Membranous prostheca, (3) Basal tooth.

Family - Mantidae

External spines of anterior tibia straight, well separated and less in number. Fore wing without eye like spiral marks.

Family - Empusidae

Antennae combed.

In India very little information is available on state wise mantid fauna, except the works of Mukherjee and Hazra, 1983; 1985, from Maharashtra and Kerala respectively. Mukherjee and Hazra (1993) also described mantid fauna of West Bengal. In past Beier (1930, 1941, 1963 & 1964), Bolivar (1897), Burmeister (1938), Giglo-Tos (1912, 1915, 1927), Kirby (1904), Lefebvre (1835), Linnaeus (1758), Mukherjee and Hazra (1983, 1985, 1993), Olivier (1792), Saussure (1869, 1870, 1871), Serville (1838), Stal (1813, 1871, 1877), Thunberg (1784), Werner (1931), West wood (1889) and Wood-Mason (1877, 1878, 1879, 1882, 1884, 1889, 1891) worked on mantids at global level. In the description of species the terminology adopted is same that of Beiet (1964).

GENUS *CREOBROTER* AUDINET - SERVILLE

Creobroter kolhapurensis sp. nov.
(Plate 35, Figs. 202-204)

Male

68 mm long, green; forewing 46 mm long and 12 mm wide, whitish green with spot; antenna brownish, 25 mm long, normal.

Head

5 mm long and 3 mm wide, brownish green, triangular, highly movable, concealed beneath the pronotum, face vertial; eyes large, greenish, 2 mm radius; ocelli three, interorbital distance 2.6 mm.

Antenna (Fig. 203)

More than 60 segmented, terminal segment small, brownish; scape longer-brownish; pedicel short; 1st segment longer than other segments.

Flagellar Formula

SL/W = 1.92, PL/W = 1.3, 1L/W = 0.55, 2L/W = 0.75, 75L/W = 4.

Mouth parts, brown, biting and chewing type; clypeus and labrum well developed; mandible strong; maxillae with distinct lobes and 5 segmented maxillary palpus.

Thorax

24 mm long and 6 mm broad, prothorax short, green, elongated, meso and meta thorax short, green; prothroax with laterally toothed margin.

Fore wings (Fig. 202)

46 mm long and 12 mm wide, white spot at middle, dark green, folded over the abdomen and sides; heteronomus; hind wings transparent whitish green, 45 mm long, elongated, membranous, folded beneath tegmina; fore wings with eye like spiral marks, eye like spot of forewing little proximally of middle.

Fore Leg (Fig. 204)

Coxa 20 mm long and 4 mm wide, greenish brown; 5 short spines; trochanter 4 mm long and 3.5 mm wide, greenish brown, somewhat triangular; femur 24 mm long and 4 mm wide, denticulate with brown, 9 long spines, 8 short spines internal spine 21 mm long; tibia 10 mm long and 1.8 mm wide, external spines of anterior tibia 23 mm and very close beet; 2 tibials spurs, unequal and one blackish brown long and one brownish with dark brown tip; tarsi 5 segmented, 6 mm long and 0.05 mm wide, tips blackish brown; claw 0.10 mm brown, curved.

Abdomen

39 mm long and 8.7 mm wide, segmented, compressed, ventrally green, dorsally yellowish, shorter than fore wings.

Female

Unknown.

Holotype

Male, India: Maharashtra, Kagal, Coll. 9-IX-2000, Vaishali J. Patil, on tobacco (*Nicotiana tabacum*) antenna, fore legs on card sheet labelled as above.

Paratype

Male 4, data same as in holotype. Coll. 9-IX-2000 from 12-XI-2002, Vaishali J. Patil, India: Maharashtra, Kolhapur.

Prey

Aphid, *A. craccivora*, grasshoppers, small moths & butterflies.

Prey Plant

Tobacco *N. tabacum*

Discussion

This species runs close to *Creobroter laevicollis* (Saussure), by having following characters:

1. Eye spot on fore wing appears little proximatly placed.
2. Coxa with 5 short spines.
3. Fore wings well developed.
4. Pronotum denticulate.

However, it differs from above species by having following characters:

1. Annal membrane not black.
2. Hind wing not pink at base.
3. No smoky patches in the centre of wing, patch round.
4. Abdomen yellowish.
5. Femur internally spines with 21 mm long and short brown spines and black at tips.
6. Flagellar Formula:
   SL/W = 1.92, PL/W = 1.3, 1L/W = 0.55, 2L/W = 0.75, 75L/W = 4.

Creobroter maharashtri sp. nov.

(Plate 35, Figs. 205, 206)

Female

Body 65 mm long, light green; fore wing 53 mm long and 10 mm wide, with eye like spot in the middle of length; antenna 20 mm long, brownish, normal.

Head

5 mm long and 7 mm wide, green, triangular, movable, concealed beneath the pronotum; vertex with a spine; eyes large whitish green

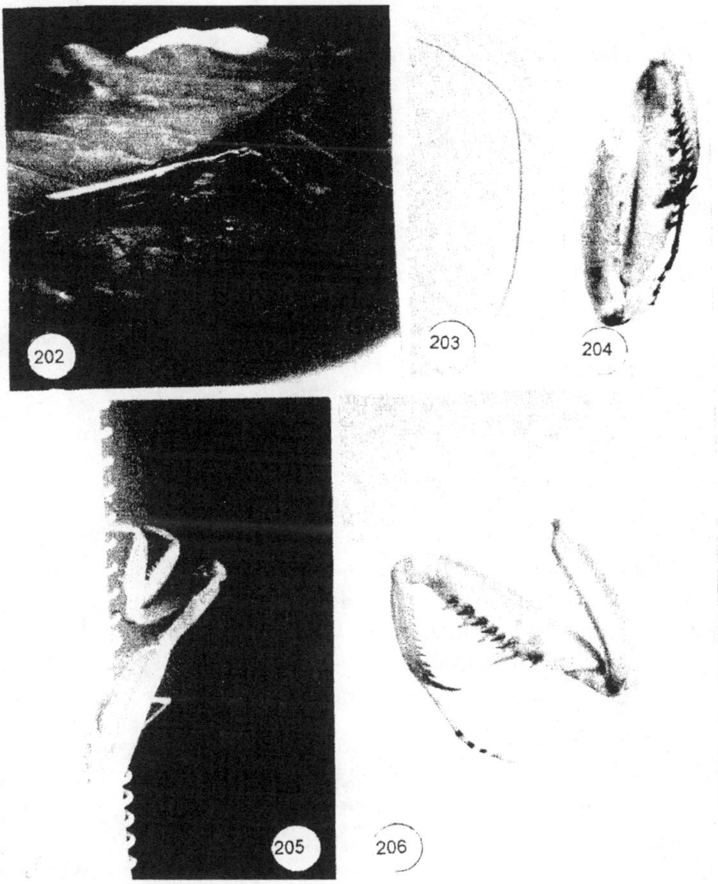

Plate XXXV

Fig. 202: ***Cerobroter kolhapurensis*** sp. nov. (Female), Fig. 203: Antenna, Fig. 204: Fore leg, Fig. 205: ***Creobroter maharashtri*** sp. nov. (Female), Fig. 206: Fore leg.

transparent, 3 mm radius, large; interorbital distance 3 mm; ocelli three.

Antenna

More than 60 segmented, 20 mm long, brownish; scape large brownish, elongated, hairy; pedicel short; 1st segment longer than other few segments; terminal segment hairy, elongated, longer, brownish.

Flagellar Formula

SL/W = 2.90, PL/W = 2.1, 1L/W = 1.93, 2 L/W = 1.40, 86 L/W = 3.

Mouth parts strong, mandibles brownish and denticulate blackish at apical tooth; maxillary palpi 5 segmented, terminal segment brownish black; maxilla brownish, 4 segmented, maxillary palpi; labrum greenish brown.

Thorax

25 mm long and 7 mm wide, greenish, prothorax short, bicarniate, denticulate, broad; meso and meta thorax smaller and equal; pronotum denticulate.

Fore Wings (Fig. 205)

With eye spot in the middle of the length, membranous, 53 mm long and 10 mm wide, whitish green. Fore wing longer than abdomen, elongated, folded over the abdomen; hind wing transparent, folded beneath tegmina, whitish transparent.

Fore Legs (Fig. 206)

Coxa 11 mm long and 3mm wide, strong, greenish white, denticulate with four spines; trochanter somewhat triangular broad, 3 mm long and 4 mm wide, greenish; femur 18 mm long and 6 mm wide, yellowish green, 22 brown spines with blackish tips; femur broad and strong; tibia 16 mm long and 5 mm wide, spines 20, greenish and dark brownish black tips, smaller in size; tarsi 8 mm long and 2 mm wide, 5 segmented; claw 0.10 mm long, bifid, brown.

Abdomen

35 mm long and 13 mm wide, 6 segmented excluding ovipositor, yellowish green, narrow at tip.

Male

Unknown.

Holotype

Female, India: Maharashtra, Kagal, Coll. 9-X-200 1, Vaishali J. Patil on Sorghum (*Sorghum vulguare* Pers.), fore leg of female labelled as above.

Paratype

Female 3, India: Maharashtra, Kolhapur CoIl. 19-1X-2000 to 4-111-2002, Vaishali J. Patil, data same as in holotype.

Prey

Jowar aphid, *Rhophalosiphum maidis* Fitch., *Helicoverpa armigera* Hubn., *Spodoptera* larvae; Mosquitoes, *Culex*, *Aedes* sp.

Prey Plant

Sorghum (*S. vulguare*) Grasses, Shrubs, Mulberry etc.

Discussion

This species runs close to *Creobroter lavevicollis* (Sss.) by having following characters:

1. Body colour, shape and size.
2. Fore wing with eye like spot in the middle of length.
3. Vertex with spines.
4. Prothorax green, pronotum short.

However, it differs from above species by having following characters:

1. Pronotum bicarniate, denticulate.
2. Coxa strong denticulate with four spines.
3. Femur yellowish green, 22 brown spines with blackish tip.
4. Tibia with 20 greenish spines and dark brown at tip.
5. Flagellar Formula:

   SL/W = 2.90, PL/W = 2.1, 1 L/W = 1.93, 2L/W = 1.40, 86L/W = 3.

Creobroter marathi sp. nov.

(Plate 36, Figs. 207 to 209)

Female

Body 63 mm long, yellowish brown; forewing 44 mm long and 9 mm wide with eye like spot; antenna 18 mm long brown.

Head

4 mm long and 6.5 mm wide, brown, triangular, highly movable, concealed beneath the pronotum, vertical face; eyes large, brownish, 2 mm radius; oceli three; interorbital distance 4 mm.

Antenna (Fig. 208)

More than 60 segmented; scape longer, elongated, pedicel shorter than scape; 1st segment longer, elongated, hairy, brownish; terminal segment elongated, longer, brownish.

Flagellar Formula

SL/W = 3.35, PL/W = 3.1, 1L/W = 3.8, 2L/W = 1.71, 59L/W = 5.8.

Mouth parts yellowish brown, biting and chewing type; clypeus and labrum well developed; mandible strong brownish; maxillae with distinct lobes; maxillary palpi 5 segmented.

Thorax

20 mm long and 7 mm wide, brown, elongated; prothroax long, meso and meta thorax with black spot, brown, prothorax with laterally toothed margin.

Fore Wing (Fig. 207)

Fore wings 44 mm long and 14 mm wide, brown with eye like spiral marks, eye like spot of forewing little proximally of middle.

Fore Legs (Fig. 209)

Coxa 10 mm long and 4 mm wide, yellowish brown, six short spines; trochanter 4 mm long and 2 mm wide, somewhat triangular; femur 14 mm long and 4 mm wide with 21 long and short brown spines, black at tip; tibia 6 mm long and 2 mm wide, 13 pairs of short, dark brown spines and one long spine with blackish brown tip; tarsus 8 mm long and 2 mm wide, 1st, 2nd and 3rd segments of tarsi with blackish spots; claw 0.2 mm long, brown and curved.

Abdomen

39 mm long and 12 wide; yellowish brown, 6 segmented, compressed shorter than forewings.

Male

Unknown.

Holotype

Female, India: Maharashtra, Kolhapur Coll. 12-XII-2001, T. V. Sathe, antenna, fore legs on card sheet labelled as above.

Prey

Mosquitoes (*Culex* sp, *Aedes* sp), scale insects, mealy bugs, aphids, grasshoppers.

Prey Plant

Custard apple (*Annona squmosa* L.)

Discussion

This species runs close to *Creobroter laevicollis* (Saussure) by having following characters:

1. Eye spot on fore wing appears a little proximately placed.
2. Body size and shape.
3. Coxa with 6 spines.
4. Fore wings well developed.
5. Pronotoum denticulate.

However, it differs from above species by having following characters:

1. Body colour yellowish brown.
2. Hind wing not pink at base.
3. Femur with 21 long spines with brown tip.
4. Tibia with 13 pairs short dark brown spines.
5. 1st, 2nd and 3rd segment of tarsi with blackish spot.
6. Thorax with large black spot.
7. Flagellar Formula:
   SL/W = 3.35, PL/W = 3.1, 1L/W = 3.8, 2L/W = 1.71, 59L/W = 5.8.

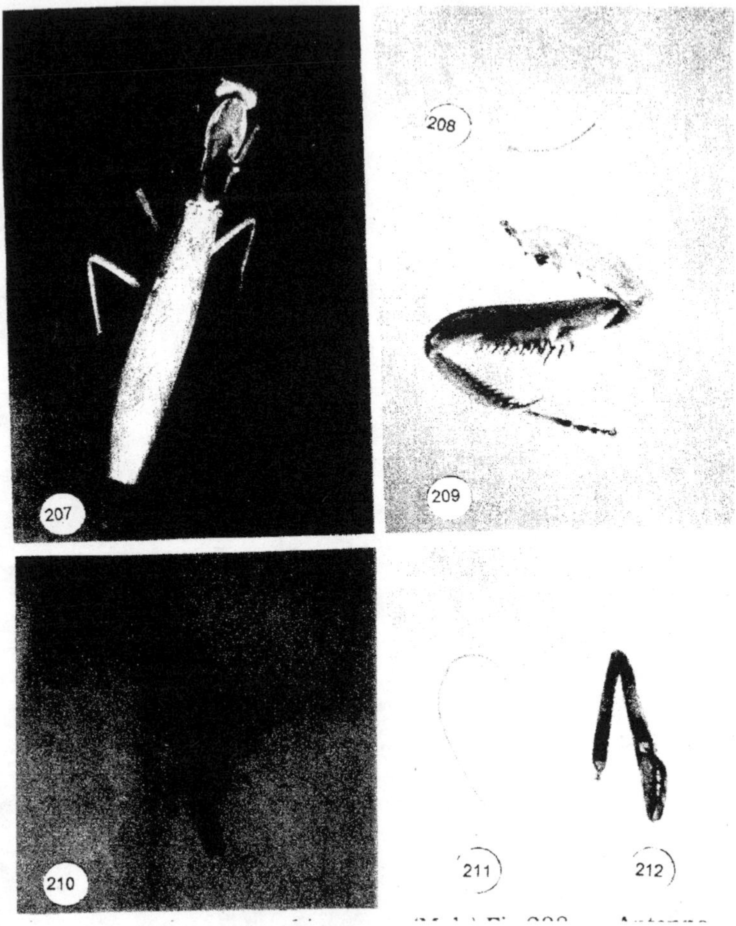

Plate XXXVI

Fig. 207: *Ceroboter marathi* sp. nov. (Male), Fig. 208: Antenna, Fig. 209: Fore leg, Fig. 210: *Humberticella mulberrae* sp. nov. (Male), Fig. 211: Antenna, Fig. 212: Fore leg.

GENUS *HUMBERTIELLA* SAUS. 1869

Humbertiella mulberae sp. nov.
(Plate 36, Figs. 210 to 212)

Male

Body brownish, small, 43 mm long; fore wing 31 mm long and 4 mm wide, brown; antenna 12 mm long, brown, normal.

Head

2 mm long and 3 mm wide, brown, small, triangular, movable; eyes brownish, small with 1.7 mm radius; ocelli three, interorbital distance 2.0 mm.

Antenna (Fig. 211)

12 mm long, 52 segmented, normal, brown, scape longer than pedicel, terminal segment dark brown.

Flagellar Formula

SL/W = 1.26, PL/W = 1.50, 1L/W = 1.98, 2L/W = 0.59, 50L/W = 1.18.

Mouth parts brownish, small, biting and chewing; clypeus and labrum small, well developed; mandible brownish, small, strong; 5 segmented maxillary palpi.

Thorax

13 mm long and 3 mm wide, blackish brown, denticulate; prothorax longer, not bicarniate, with black dots; meso and meta thorax short.

Fore Wing (Fig. 210)

Forewing brown, longer than abdomen, 31 mm long and 4 mm wide, some black spots on lateral sides, not transparent, folded over the abdomen and sides; hind wing longer than forewing, folded beneath tegmina.

Fore Leg (Fig. 212)

Coxa 8 mm long and 2 mm wide, brown, 4 small spines; trochancter 2 mm long and 1 mm wide; femur 10 mm long and 2 mm wide, 10 brown spines and one long brown spine; tibia 4 mm long and 1 mm wide, 12 black spines, 6 brown spines with black tip, one long brown, tibia shows blackish edges at inner side, anterior tibia denticulate; tarsi 4 mm long; claw 0.10 mm long, brown.

Abdomen

28 mm long and 3 mm wide, brownish, shorter than fore wings, not flat.

Female

Unknown.

Holotype

Male, India: Maharashtra, Kolhapur, Coll. 20-VIII-2000, T.V. Sathe, antenna, fore leg of male on card sheet labelled as above.

Prey

White fly, jassids, soft scales, mealy bugs, *Spodopteran* caterpillar, *Sylepta derogata* caterpillar, butterflies.

Prey Plant

Mulberry, *Morus alba* (Local variety).

Discussion

This species runs close to *Humbertiella indica* (Sss) by having following characters:

1. Body size, shape and length.
2. Fore wings length and size.
3. Thorax.

However, this species differs, from above species by having following characters:

1. Fore wings brown with black dots at lateral sides.
2. Prothorax not bicarniate with black dots.
3. Body colour brown.
4. Coxa brown with 4 small brown spines.
5. Femur with 10 brown spines and one long brown spine.
6. Tibia with 12 black spines and 6 brown spines with black tip, tibia shows blackish edges at inner side, anterior tibia denticulate.
7. Flagellar Formula:

   SL/W = 1.26, PL/W = 1.50, 1L/W = 1.98, 2L/W = 0.59, 50 L/W = 1.18.

GENUS *EREMOPLANA* WLK.

Eremoplana elongata sp. nov.
(Plate 37, Figs. 213 to 215)

Female

80 mm long excluding ovipositor, brown, long; forewing 47 mm long and 7 mm broad; antenna 20 mm long brownish, normal.

Head

1 mm long and 3 mm wide, brown, small, triangular, highly moveable, never concealed beneath the pronotum, face vertical; eyes large, brown, 2 mm radius; ocelli three; interorbital distance 2.2 mm.

Antenna (Fig. 214)

20 mm long, normal, more than 52 segmented including scape and pedicel, dark brown terminal segment.

Flagellar Formula

SL/W = 1.6, PL/W = 1.71, 1L/W = 2, 2L/W = 0.7, 50L/W = 1.28.

Mouth parts, brown, biting and chewing, clypeus and labrum well developed; mandibles strong; maxillae with distinct lobes and maxillary palpi 5 segmented; labium fused with stipes; triarticulate maxillary palpi.

Thorax

23 mm long and 2 mm broad, dark brown; prothorax elongated brown with small spines, freely movable, laterally with toothed margins; meso and metathorax equal, short, immovable.

Fore Wing (Fig. 213)

Fore wings 47 mm long and 7 mm broad, shorter than abdomen, wavy marking on anterior side, folded, flat over the abdomen and sides; heteronomous; tegmina leathery; hind wing longer, 50 mm long and 6 mm wide, membranous, folded beneath tegmina in repose.

Fore Leg (Fig. 215)

Coxa 11 mm long and 2 mm wide, brownish black, denticulate; trochanter pointed at one side, 1.8 mm and 0.9 mm wide; femur 16 mm long and 1.7 mm wide, large, with row of strong 7 sharp spines and many small unequal spines; tibia 11 mm long and 0.8 mm wide, with 10 black spines; tarsi 5 segmented, 6 mm long and 0.5 mm wide; claw 0.12 mm long, brown, curved.

Middle Leg and Hind Leg

Coxa smaller than fore leg's coxa, anterior femur with flux, flux at middle in tibia with 2 small spines.

Abdomen

56 mm long excluding ovipositor and 4 mm broad, 6 segmented, compressed, brown, longer than fore wings, and cerci flat.

Male

Unknown.

Holotype

Female, India: Maharashtra, Kolhapur, Coll. 8-XI-2000, Vaishali J. Patil, antenna, fore legs of female on card sheet labelled as above.

Prey

Grasshoppers, mealy bugs and aphids.

Prey Plant

Grasses, custard apple (*A. sqamosa*).

Discussion

This species runs close to *Eremoplana microptera* Wlk. by having following characters:

1. Body colour, length and side.
2. Prothorax.
3. Length of abdomen and forewings.

However, it differs from above species by having following characters:

1. Fore leg femur with 7 sharp spines and many unequal small spines; tibia with 10 black spines.
2. Fore wing with wavy marking at anterior side.
3. Middle and hind legs with flux on femur and tibia.
4. Anal cerci flat.
5. Flagellar Formula:

   SL/W = 1.6, PL/W = 1.71, 1L/W = 2, 2L/W = 0.7, 50L/W = 1.28.

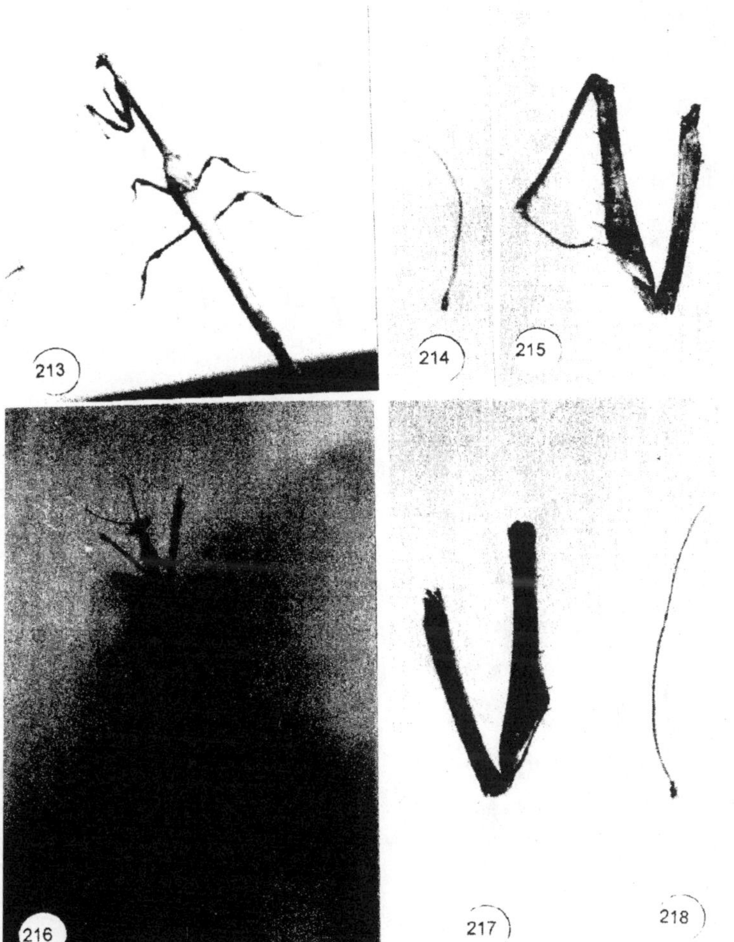

Plate XXXVII

Fig. 213: *Eremoplana elongata* sp. nov. (Female), Fig. 214: Antenna, Fig. 215: Fore leg, Fig. 216: *Deiphobe kolhapurensis* sp. nov. (Female), Fig. 217: Fore leg, Fig. 218: Antenna.

GENUS *DEIPHOBE* KIRBY 1904

Deiphobe kolhapurensis sp. nov.

(Plate 37, Figs. 216 to 218)

Female

Body dark blackish, 82 mm long; forewing 38 mm long and 6 mm wide, brown, with blackish spots; antenna blackish brown, normal, 13 mm long.

Head

2 mm long and 4 mm wide, blackish brown, small triangular, movable; eyes blackish with pointed horn like; ocelli three; interorbital distance 1.8 mm, eyes with 2.5 mm radius.

Antenna (Fig. 218)

More than 52 segmented, normal, blackish brown, 13 mm long including scape and pedicel, dark brown terminal segment.

Flagellar Formula

SL/W = 1.2, PL/W = 1.68, 1L/W = 2, 2L/W = 0.68,50 L/W = 1.21.

Mouth parts blackish brown, biting and chewing; clypeus and labrum well developed; mandible strong; maxillary palpi 5 segmented.

Thorax

23 mm long and 2.5 mm wide, blackish brown, denticulate, prothorax bicarniate, elongated with small spines, meso and meta thorax short.

Fore Wing (Fig. 216)

Fore wings 38 mm long and 6 mm wide, shorter than abdomen, having black patches/spots, transparent, folded over the abdomen and sides; heteronomous; hind wing longer than fore wing, folded beneath tegmina.

Fore Leg (Fig. 217)

Coxa 11 mm long and 2 mm wide; denticulate with small brown spines; trochanter 1.8 mm long and 0.9 mm wide, somewhat denticulate, triangular; femur 16 mm long and 1.7 mm wide, 10 brown spines and one brown long spine; tibia 11 mm long and 0.8

mm wide, with 12 black spines, 6 brown spines with black tip and one long brown spine, tibia shows blackish edges at inner side, anterior tibia denticulate; tarsi 6 mm long, 5 segmented, blackish brown with bifid claw.

Abdomen

57 mm long and 5 mm wide, 6 segmented blackish, longer than forewings, flat, slender, anal cerci flat, brownish.

Male

Unknown.

Holotype

Female, India: Maharashtra, Kolhapur, Coll. 12-XII-2001, T.V. Sathe, antenna, fore leg of female on card sheet labelled as above.

Prey

Grasshopper, caterpillars, psyllid, aphids, white fly.

Prey Plant

Guava (*Psidium guajava* L.).

Discussion

This species runs close to *Deiphobe incisa* by having following characters:

1. Body size and shape.
2. Abdominal length.
3. Fore wing length.

However, it differs from above species by having following characters:

1. Eyes blackish with pointed horn like.
2. Fore wings with blackish spots.
3. Prothorax bicarniate, denticulate, elongated with small spines.
4. Fore leg, coxa denticulate with small brown spines; trochancter somewhat denticulate; femur with 10 brown spines and one long brown spine; tibia with 6 brown spines with black tip, tibia shows blackish edges at inner side.

5. Anal cerci flat, brownish.
6. Flagellar Formula:
   SL/W = 1.2, PL/W = 1.68, 1L/W = 2, 2L/W = 0.68, 50L/W = 1.21.

GENUS *HIERODULA* SAUS. 1864

Hierodula orientalis Sp. Nov.

(Plate 38, Figs. 219 to 221)

Female

85 mm long, pale green; fore wing 55 mm long and 14 mm wide, without spot; antenna 20 mm long, brown.

Head

4 mm long and 6 mm wide, greenish yellowish, triangular, highly movable, concealed beneath the pronotum, face vertical; eyes large, greenish, 2.1 mm radius; ocelli three; interorbital distance 3.8 mm.

Antenna (Fig. 220)

More than 60 segmented; scape longer, elongated; pedicel, shorter than scape; flagellar 1st segment longer, elongated, hairy, brownish; terminal segment elongated, longer, brownish.

Flagellar Formula

SL/W = 3.25, PL/W = 3, 1L/W = 3, 2L/W = 1.66, 59L/W = 5.5.

Mouth parts yellowish brown, biting and chewing type; clypeus and labrum well developed; mandibles strong, brownish; maxillae with distinct lobes and maxillary palpi 5 segmented.

Thorax

27 mm long and 7 mm wide, green with light brown edges, elongated; prothorax green, long pronotum dilation, does not extend upto base; short, little narrowed posteriorly, meso and meta thorax short, green; prothorax without laterally toothed margin.

Fore Wing (Fig. 219)

Fore wing 55 mm long and 14 mm wide, pale green, elongated narrow at tip and brownish; without white spot; folded over the abdomen and sides; hind wing transparent greenish, 57 mm long and 7 mm wide, membranous, folded beneath tegmina.

Fore Legs (Fig. 221)

Coxa 18 mm long and 3 mm wide, green, elongated with small yellow spots, anterior coxa with whitish patches at bases of smaller spines; trochancter 3.8 mm long and 2.5 mm wide, somewhat triangular; femur greenish brown, 25 mm long and 4 mm wide with 23 brownish spines and longer brownish black spines; elongated, broad at middle; tibia 12 mm long and 2.1 mm wide with long 18 brownish black spines; tibial spurs brownish; tarsi 8 mm long and 0.6 mm wide, brown; claw 0.13 mm long, brown, curved.

Abdomen

54 mm long and 10 mm broad, 6 segmented, greenish brown.

Male

Unknown.

Holotype

Female, India: Maharashtra, Kolhapur, Coll. 15-XII-2001, Vaishali J. Patil, antenna, fore legs of female on card sheet labelled as above.

Paratype

Female: 2, data same as in holotype, Coll. from 15-XII-2001 to 20-X-2002, Vaishali J. Patil, India, Maharashtra, Kolhapur.

Prey

Grasshoppers, cockroaches, caterpillars, *H. armigera, E. fabia,* Mosquitoes.

Prey Plant

Cotton (*Gossypium hirsutum* L.); custard apple (*A. sqamosa*), mulberry (*M. alba*).

Discussion

This species runs close to *Hierodula* (*Hierodula*) *bipapilla* (Serville-Audinet) by having following characters:

1. Anterior coxa with whitish patches at bases of smaller spines.
2. Internal spines of femur black.
3. Pronotum short, little narrowed posteriorly.
4. Fore hind wings longer than abdomen.
5. Dilation of pronotum does not extend upto base.

However, it differs from the above species by having following characters:

1. Metazona of prosternum without blackish bands.
2. Body size.
3. Anterior femur with small brown spines.
4. Internal tip of tarsi not black.
5. Antenna thin and yellowish brown.
6. Tibial spurs brownish.
7. Flagellar Formula:
   SL/W = 3.25, PL/W = 3, 1L/W = 3, 2L/W = 1.66, 59L/W = 5.5.

Hierodula shivajiensis sp. nov.
(Plate 38, Figs. 222, 223)

Male

Body 60 mm long, yellowish; fore wing 40 mm long and 6 mm wide, without spot; antenna 17 mm long, yellowish.

Head

3 mm long and 4 mm wide, yellowish, triangular, highly movable, concealed beneath the pronotum, vertical face; eyes large, brown, 2.2 mm radius; ocelli three, yellow; interorbital distance 2.5 mm.

Antenna

More than 60 segmented, 15 mm long; scape large, yellowish brown, elongated; pedicel short; flagellar 1st segment longer than others, elongated, hairy, yellowish; terminal segment elongated longer, yellowish.

Flagellar Formula

SL/W = 2.98, PL/W = 2.40, 1L/W = 2.42, 2L/W = 1.53, 59 L/W = 5.1.

Mouth parts yellowish brown, biting and chewing type; clypeus and labrum well developed; strong mandible; distinct maxillae lobes and 5 segmented maxillary palpi.

Thorax

18 mm long and 5 mm wide, yellowish, elongated, prothorax

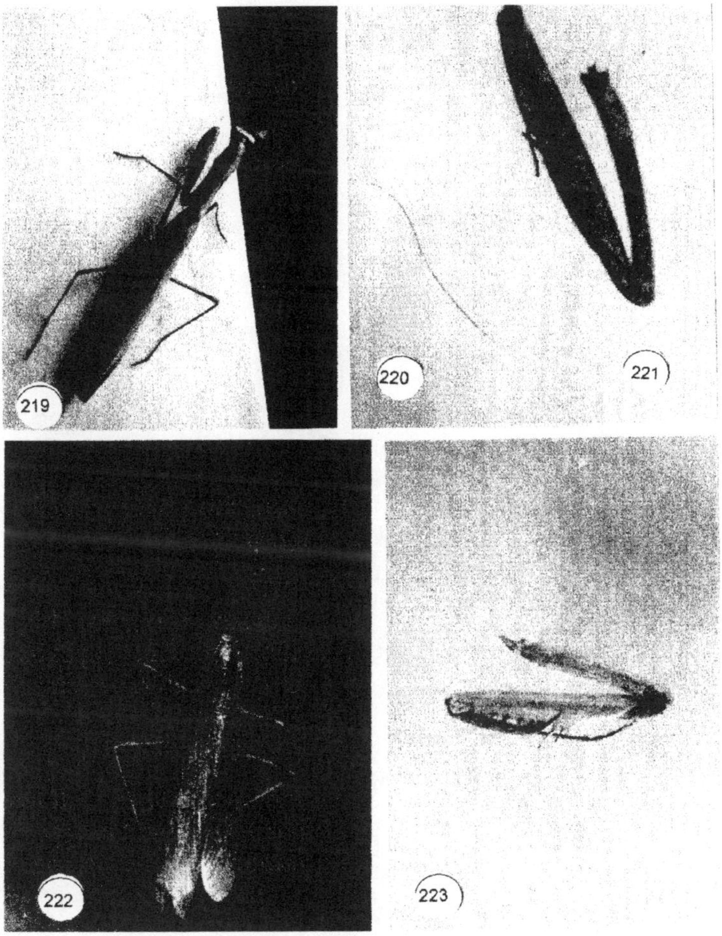

Plate XXXVIII

Fig. 219: *Hierodula orientalis* sp. nov. (Female), Fig. 220: Antenna, Fig. 221: Fore leg, Fig. 222: *Hierodula shivajiensis* sp. nov. (Male), Fig. 223: Fore leg.

short yellowish, meso and meta thorax short and equal, yellowish; pronotum short, prothorasic lateral toothed margin absent.

Fore Wing (Fig. 222)

Forewing membranous, 40 mm long and 6 mm wide, yellowish, transparent, without spot, longer than abdomen, elongated, narrow at tip and folded over the abdomen and sides; hind wing transparent 44 mm long and 7 mm wide, folded beneath tegmina.

Fore Leg (Fig. 223)

Coxa 10 mm and 2 mm wide, with very small spines, whitish small spots present on coxa, with 3-4 verrucose with few spinules among them; trochancter 3 mm long and 2 mm wide somewhat triangular, yellowish; femur 14 mm long and 3 mm wide, 5 internal blackish spines, 14 yellowish spines with black tip and only one long yellowish spine; tibia 6 mm long and 1 mm wide, 11 small yellowish spines and one long black spine; tarsi 7 mm long and 0.90 mm wide; 1st segment longer than other segments, elongated; 2nd, 3rd and 4th segments brownish black; last terminal segment terminated into brownish bifid claw; claw 0.2 mm, brown, curved.

Abdomen

39 mm long and 10 mm wide, yellowish brown.

Female

Unknown.

Holotype

Male, India: Maharashtra, Kolhapur. Coll. 21-VI-2000, T.V. Sathe, fore legs of male on card sheet labelled as above.

Prey

Jassids, mealy bugs, white files, aphids, caterpillars of *S. litura, H. amrigera, Mythimna separata* Walker.

Prey Plant

Mulberry (*M. alba*), Jowar (*S. vulgare*).

Discussion

This species runs to *Hierodula bipapilla* (Serulle-Audinct) by having following characters:

1. Body size and shape.

2. Eyes.
3. Pronotum short, little narrowed posteriorly.
4. Coxa with few spinules among them.
5. Fore hind wings longer than abdomen.

However, it differs from above species by having following characters:

1. Body colour brown.
2. Frontal scelerite not bicarinate.
3. Metazona of prosternum not with two less distinct blackish bands.
4. Femur with five black spines and 14 yellowish spines with black tip and only one long spine.
5. Tibia with 11 small yellowish spines and one black spine.
6. Flagellar Formula:
   SL/W = 2.98, PL/W = 2.40, 1L/W = 2.42, 2L/W = 1.53, 59 L/W = 5.1.

GENUS *SCHIZOCEPHALA* LINNAEUS

Schizocephala gramminae sp. nov.

(Plate 39, Figs. 224 to 226)

Female

Body large, slender, stick like, 95 mm long and 3 mm wide, dark green; wings short; antenna thick, 30 mm long, yellowish brown, combed.

Head

6 mm long and 3 mm wide, green, triangular, small, movable, face vertical; head concealed beaneath the pronotum; eyes elongated conically, brownish green, 1.8 mm radius; ocelli three; interobital distance 2.5 mm.

Antenna (Fig. 225)

30 mm long, yellowish brown, combed, many segmented, posterior dark brown, anterior yellowish, broader at base and thin at apex.

Flagellar Formula

SL/W = 2.7, PL/W = 1.04, 1L/W = 2.42, 2L/W = 2.1, 86L/W = 3.

Mouth parts yellowish brown, biting and chewing type; clypeus and labrum well developed; mandible strong; maxilla with distinct lobes and 5 segmented maxillary palpi.

Thorax

41 mm long and 2 mm wide, green, long, prothorax elongated, laterally without toothed marking; meso and metathorax long; 10 black spots on ventral side of thorax.

Fore Wing (Fig. 224)

Fore wing well developed, shorter than abdomen, pale green, narrow, hind wing transparent.

Fore Leg (Fig. 226)

Spines of fore leg black at tip only; coxa 15 mm long and 3 mm wide, greenish brown, straight, elongated; trochanter 2 mm long and 0.7 mm wide, greenish; femur 24 mm long with only one spine long, and 6 short spines externally, 5th one strong, brown; hind and middle tibiae ventrally not spinous, greenish; tibia 5 mm long and 0.5 mm wide, yellowish green with small spines; tarsi yellowish, 3 mm long and 0.3 mm wide; claw yellowish.

Abdomen

48 mm long and 4 mm wide, dark green, flat, last segment and appendages brownish.

Male

Unknown.

Holotype

Female, India: Maharashtra, Kolhapur, Coll. 15-VIII-2001, Vaishali J. Patil, antenna, fore legs of female on card sheet labelled as above.

Paratype

Female, 1; data same as in holotype, Coll. 10-X-2001, Vaishali J. Patil, India, Maharashtra.

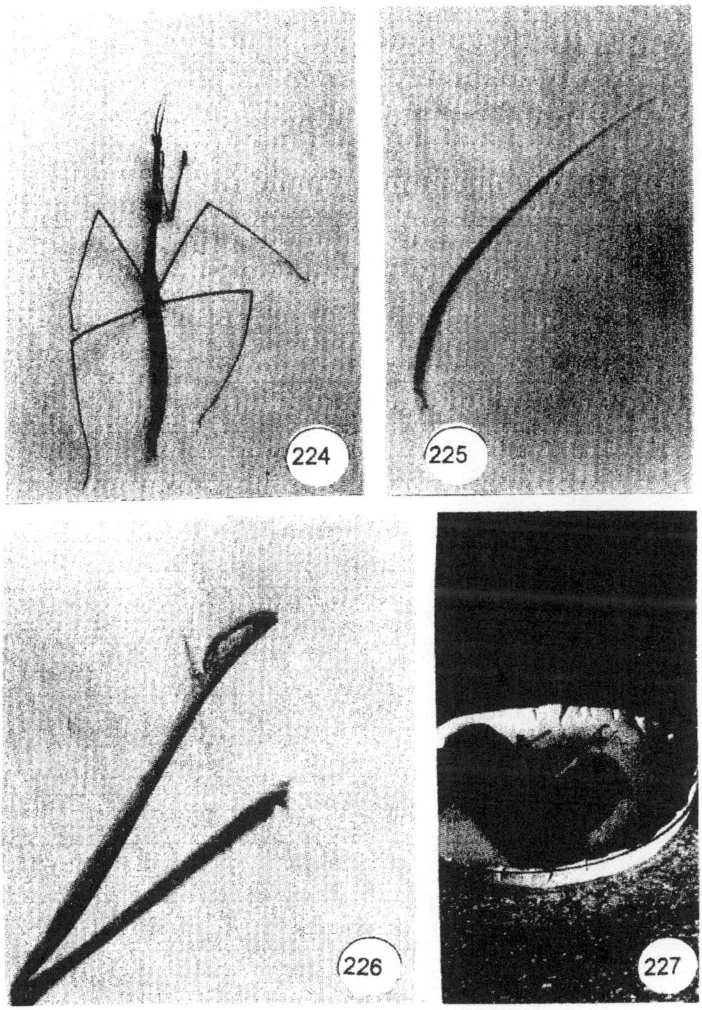

Plate XXXIX

Fig. 224: ***Schizocephala gramminae*** sp. nov. (Female), Fig. 225: Antenna, Fig. 226: Fore leg, Fig. 227: *Hierodula* sp. (Brown).

Prey

Soft bodied homopterous insect pests, such as jassids, aphids, etc.

Prey Plant

Grasses.

Discussion

This species runs close to *Schizocephala bicornis* (Linnaeus) by having following characters:

1. Body green in colour.
2. Spines of fore leg black at tip only.
3. Femur with 6 external spines.
4. Fore wings well developed.

However, it differs from above species by having following characters:

1. Antennal base not thicker.
2. Middle and hind tibiae ventrally non spinous.
3. Ten black small spots on ventral side of thorax.
4. Femur with internally 11 spines.
5. Flagellar Formula:
   SL/W = 2.7, PL/W = 1.04, 1L/W = 2.42, 2L/W = 2.1, 88L/W = 3.

Chapter 6
Biology of Predators

Biology plays an important role in mass rearing of pest predators and their utilization in pest management programme. The success or failure of biological pest control programme is depended on biological peculiarities of predators. However, very little is known about the biological aspects of predaceous insects in India.

The structure and number of larval instars has great importance in identification of the species. General criteria for taxonomic assessment is the adult characteristics of an individual. However, biology can also be reliable source for identification of the species. There occur changes in the pattern of the characters in all individuals during embryonic development. There are various animal groups where the classification is usefully aided by the characters of their immature stages. Thus, biology and biometry of immature forms help in correlating the probable phylogeny of the groups.

Coccinellid Beetles

The review of literature indicates that Kapur (1950), Hodek (1973), Shrivastava *et al.* (1978), Singh & Malhotra (1979), Rao *et al.* (1981), Diraviam & Viraktamath (1991), Patro & Behra (1992), Rawat *et al.* (1992), Omkar and Bind (1993), Sathe & Bhosale (2001), etc. have contributed on biological and biometrical aspects of coccinellids from India.

Coccinellids dominantly predate on aphids than the insects of any other group. Both the larvae and adults are voracious feeders of aphids. They can play an important role in the biological control of homopterous pests.

The life cycle of a typical coccineid beetle consists of egg, four/ five larval instars (mostly four), prepupa, pupa and adult. The eggs are oval or cigar shaped, smooth, whitish yellowish or orange or brownish and laid in clusters or singly on prey host plant preferably on underside of leaves or in cracks and crevices. Each cluster may contain 5-50 eggs. The larvae of coccinellids are of mainly following types:

1. Elongate fusiform: ex-Coccinellini tribe,
2. Cylindrical oval and more or less convex: ex-chilocorini and epilachnini,
3. Fusiform with wax like exudation: ex-Scymnini,
4. Elliptical and moderately convex: ex-Noviini,
5. Short oval and strongly flattened: ex-Platynaspini and
6. Fusiform and elongate oval: ex-Sticholotinids

The full grown grubs of coccinellid beetles (Fig. 82) usually range between 1.2 mm to 24.00 mm in their body length. Their dorsal surface is covered by cottony/waxy material or some times without such material. The larvae are usually armed with well-sclerotized plates or tubercles when they possess no waxy material. The head capsule of a larva is usually rectangular and broader than long but, exceptionally may be elongated. Antennae of a typical form consists 3 segments and are short and located near the mandibles. Antenna shows 1-3 segments. The mandibles are always subtriangular. The apex of mandible is usually bifid or simple but, never multidenticulate. The maxillary palpus contain 2-3 segments and labial palpus is always with a very small and usually with 2 segments.

The legs of a coccinellid grub consists of coxa, trochanter, femur and tibiotarsus (tibia). The thorax shows well defined three segments. The prothorax is transverse oval while, meso and metathorax are strongly transverse and distinctly broader than the prothorax. The prothorax has sclerotized plates divided at middle line. The plate some times shows 1-5 setose projections. The mesothorax is demarketed by dorsal, lateral and ventral areas. The dorsal area provides two pairs of sclerotized plates or tuburcles, often they may be fused in a large plate. The mesothoracic lateral area is divided into anterior and posterior parts and armed by long projection while, mesothoracic ventral area provides two pairs of indistinct sclerotizations. The metathorax and mesothorax are more or less equal in length and structure. Some times mesothorax may be broader than metathorax.

The abdomen of a coccinellid grub consists of ten segments. The 9th abdominal segment is usually simple and without urogomphi. Each abdominal segment is demarketed by four areas namely, dorsal, dorsolateral, lateral, ventrolateral and ventral. Each area has a single sclerotized tuburcle. The 10th segment usually serves as a sucking disk and helps in locomotion.

Generally, full grown grubs pupate near the point where they lived. However, pupation may also be takes place away from the living place of larvae. Large aggregations of pupae of some species often observed on the host plant of the prey.

There are two types of pupae in predaceous coccinellids. In one case, the exvium is remained at the caudal tip and in another case it is nearly intact as noticed in tribe Noviini and Chilochorini.

Newly emerged pair of adults mate within a day or two. Females have preoviposition period of about a week. Either the female oviposits within a week or enter the dormant season. Inspite of strong disparsion activity, the adult coccinellids usually stay for some time on the plants where their larval stages have been spent and then flew away in search of food/mate/shelter.

A comparative account of life cycle of three coccinellids namely, *Illeis darbari* S. & B., *Menochilus sexmaculatus* (Fab.) and *Coccinella yamini* S. & B. predating on jowar and groundnut aphids is represented in Table 13.

Praying Mantids

Praying mantids are very potential predators of agricultural pests. They are reported from various parts of the world such as India, North America, Europe, United States etc.

Praying mantids are medium to large sized, having elongated prothorax and large spined grasping and raptorial front legs. The middle and hind legs are usually slender. The mantids may be long-winged, short winged or completely wingless. When wings present, folded over the abdomen and sides called, heteronomous. Many are green, brown or mottled. Some species have brighter coloured. Their mouth parts are of biting and chewing type. Their body is slender, elongate but depressed, often leaf-like. Head is small, triangular, highly movable.

Table 13: Life Cycle of Some Coccinelid Beetles

Sl. No.	*I. darbari*	*M. sexmaculatus*	*C. yamini*
Egg	Incubation period: 3-4 days	Incubation period: 3-4 days	Incubation period: 3-4 days
Larva	Elongate fusiform	Elongate fusiform	Elongate fusiform
Ist Instar	Ist instar lasts for 5 days	This stage lasts for 3 days	This stage lasts for 24 hr.
IInd Instar	This stage lasts for 4 days	This stage lasts for one day	This stage lasts for one day only
IIIrd Instar	Duration 3 days	This stage lasts for two days	This stage lasts for two days
IVth Instar	This stage lasts for two days	This stage lasts for 5-6 days	This stage lasts for 6 days
Pupa	The pupal stage lasts for 4 days	This stage lasts for 4-5 days	The pupal stage lasts for 4-5 days.
Adult	Female: Oval, pale yellow, dorsum weekly convex, ventrum yellow, spermathica broadly, 'U' shaped (35 × 10 mm)	Female: Medium sized, elongate yellowish, 5.2 mm long, 4.0 mm broad, elytra with 3 black zigzag spots. Spermathica 0.58 mm long.	Female: Oval, reddish with black spots, 5.75 mm long, 4.40 mm broad, pronotum black, with 3½ black spots, Ventral side black, Spermathica 40 × 1.0 mm, 'U' shaped
Life cycle	Completed in 21-22 days	Completed in 18-21 days	Completed in 22-24 days
Adult longevity	22-36 days	20-34 days	24-38 days
Copulation period	113±1.76 min.	102±1.15 min	132±1.62 min
Oviposition period	8 to 15 days	7 to 15 days	8 to 17 days
Prey	*R. maidis*	*R. maidis*	*A. craccivora*

Antennae are long, setaceous. Compound eyes are large. Ocelli are three in number or may be absent.

Egg

The eggs are deposited in ootheca (Figs. 27 & 93) which is two or more chambered or rows and found fixed to some plant. The ootheca is made of whipped and solidified foam from a viscid secretion. Ootheca is an ideal thermoinsulator for the eggs. The winter is passed in the egg stage. The number of ootheca formed increased in late winter or early spring. The eggs are frequently parasitized by some Hymenopterous parasitoids such as Chalcids, Eulophids, etc. Each egg case/ootheaca may contain 50-400 eggs. Egg hatching takes place within 1-4 weeks.

Nymph

The young nymphs (Fig. 90) are well capable for predation. They have cannibolism among them. They moult 3-12 times and become adults in about a year. Praying mantids complete their life cycle within one year.

The mantids are mostly arboreal but some apterous forms are found on the ground. In general, both adults and nymphs are terrestrial. They are entirely carnivorus and lead predaceous life both as young and adult. The mantids become cautiously and perfectly motionless during watching the preys for predation. The prey is caught with the help of raptorial fore legs and then eaten.

The food of mantids comprises various kinds of insects such as flies, leafhoppers, aphids, scales, crickets, mosquitoes, grasshoppers, caterpillars, butterflies, moths, cockroaches, bees, wasps, wooly aphids on sugarcane, small lizards, frogs, etc.

Mantids are associated with plants such as grasses, bushes, trees, agricultural crops, garden plants etc. They have their own territory. The size of territory depends on availability of prey fauna. They spend a single season on one or two bushes when abundant food is available. Two or three individuals may live in close vicinity. According to some workers, a garden of size 20′ × 40′ will provide good place for 2 to 3 mantids. Praying mantids mostly visualized at morning or late afternoon, basking in the sunshine. In mantids grooming behaviour is very prominent. They clean their antennae and eyes through grooming behaviour. A mantid can consume entire body of cricket or grasshopper within 90-100 seconds.

Tiger Beetles

Tiger beetles occurs through out the world, mostly in tropical and subtropical areas of the world. They have been reported from India, Europe, North America, United States etc. Most are restricted to environments with access to open substrate including stream edges, saline flats, seashores and sand dunes. Tiger beetles are an important component of the ecosystem. They are important link in the food chain.

The tiger beetles (Fig. 25) are predacious insects mostly active at day and occurring on open substrates where they forage on several insects and small arthropods. Tiger beetles are well-known for their quick running and agile flying abilities. They are of various size, colouration and marking on their wings. They are often brightly coloured, green, brown, black with spots or bands of white. The majority of tiger beetles are ranging from one half to an inch long. Head of a tiger beetle is short and thick constricted behind the eye into a neck. Eyes are prominent, antennae moderately long, curved mandibles projecting infront of the head, mouth parts are well suited for predaceous habit. The prothorax is large and cylindrical. Eyltra is usually smooth or only finely pitted. Tiger beetles may be winged or wingless. The legs are long, slender and suited for fast running and capturing the preys.

The life history pattern of most of the beetles is more or less same. The features of life cycle of tiger beetles are given below.

Egg

Oviposition is usually initiated by female touching the substrate of ovipositing site with her antennae. Eggs are laid in soil. The female excavates a small hole up to 1 cm deep in soil, deposits a single egg, then covers the hole. Eggs are laid in late spring or early summer. Embryological development takes place in 9-29 days. Egg chorion is transparent hence helpful for noting the observations on embryological events. In certain *Cicindela* sp. incubation occurred within 10-12 days at laboratory conditions.

Larva (Grub)

In *Cicindela* sp. there are three larval instars. Larval period lasts for 1-4 years. However, most of the individuals complete their larval period within 2 years. The first instar takes 1-3 weeks the second instar lasts for 1-2 months and the third instar extends for several months to several years in field condition. Under laboratory condition and faid ad libetum the tiger beetles can complete their development from egg to adult within 60 days.

Pupa

Full grown 3rd instar grubs plug the tunnel entrance and digout a chamber for pupation within few minutes. Pupation takes place in a pupal cell. In *Cicindela* sp. pupation period lasts for 18-22 days. While in others like *Pseudoxycheila* sp. pupal period lasts for 30 days.

The majority of tiger beetles are found in damp places, in rice fields or thick vegetation, on river banks, on the seashore; some are found only on trees in forest localities, also on grassland and open ecosystem. Some species of tiger beetles begin reproduction soon after emergence while others delay reproduction until the spring after their emergence as an adult. Males are polygamous. Interspecific hiberdization is recorded in the group of tiger beetles. Most tiger beetles are diurnal. They regularly get attracted to light at night.

Tiger beetles acts as biocontrol agents of several insect pests. They feed upon grasshoppers, locusts, leafhoopers, small insects, caterpillars, rice bugs, snails, gypsy moth, brown tail moth, grubs of Rhinocerous beetle, etc.. They are predated by robber flies, lizards and birds. The tiger beetles are potential bicontrol agents of insect pests of agriculture, horticulture and forest.

Chapter 7

Pest Predator Index

Identification and making the index of biocontrol agents are important steps in biological pest control programme. Index of pest and biocontrol agents is essential for understanding the natural enemy complex of pest species and upto certain extent the ecology of the target species. Pest biocontrol index provides immediate information on the natural enemies of the pest in the specific region. Hence, the present topic is devoted for preparation of pest predator index of Kolhapur district. The pest-predator index of Kolhapur district is given in Table 14. The index have been prepared by consulting previous records and new records added during the present study.

Table 14: Pest Predator Index from Kolhapur District

Sr. No.	*Predators*	*Pests*	*Host Plant*
[A]	**ORDER- COLEOPTERA**		
(a)	**Coccinellid beetles**		
1.	*Anisolemnia dilatata* (Fab.)	*Oregma bambusicola*	Bamboo, *Bambusa bambu*
2.	*Brumoides kolhapurensis* (S.& B.)	Aphid *Dactynotus compositae* (Theobald)	Safflower, *Canthamus tinctorium* (L.) Sunflower, *Helianthus annus* L.
3.	*B. sutrualis* (Fab.)	*Phenacoccus insolitus* *Pseudococcus* sp.	Red gram, *Cajanus cajana* M.

Contd....

Table 14–contd....

Sr. No.	Predators	Pests	Host Plant
4.	*Caria dilatata* (Fab.)	*Oregama* sp.	Bamboo, *B. bambu*
5.	*Chilocorus bijugus* (Mul.)	*Coccus* sp.	Mulberry, *Morus alba* (Linn.)
6.	*C. circumdactus* (Sch.)	*Coccus* sp.	Areca *Areca catechu*
7.	*C. nigritus* (Fab.)	*Aspidoitus* sp.	Guava, *Psidium guajava* L., *Sapota, Achras zapota*, Mango *Mangifera indica* L.
		Coccus viridis (Green)	-
		Hemiberlesia lantaniae	-
8.	*C. suryaphuli* (S. & B.)	Aphid, *D. compositae*	Sunflower *H. annus*
9.	*Coccinella arcuate* (Fab.)	The groundnut aphid, *Aphis craccivora* (Koch.) *A. rumicis*	Groundnut *Arachis hypogea* L.
10.	*C. bhuimungi* (S. & B.)	The groundnut aphid, *A. craccivora*	Groundnut *A. hypogaea*
11.	*C. madhuri* (S. & B.)	The groundnut aphid, *A. craccivora*	Groundnut, *A. hypogaea*
12.	*B. repanda* (Thunburg)	The groundnut aphid, *A. craccivora*	Groundnut, *A. hypogaea*
13.	*Coccinela, C. septampunctata* (L.)	Sorghum aphid *Rhophalosiphum maidis* (Fitc.)	Jowar, *Sorghum vulgare* Pers. Maize, *Zea mays*
		Cotton aphid, *Aphis gossypii* (Clover)	Linn. Cotton, *Gossypium hirsutum* L.
14.	*C. transversalis* (Fab.)	Groundnut aphid, *A. craccivora*, Cotton aphid, *A. gossypii* *Therioaphis ononidis*	Mung, *Phaseolus mungo* L.

Contd....

Table 14–contd....

Sr. No.	Predators	Pests	Host Plant
15.	*C. yamini* (S. & B.)	Cotton aphid, *A. gossypii,* Groundnut aphid, *A. craccivora*	Cotton, *G. hirsutum* Groundnut, *A. hypogaea*
16.	*C. yogini* (S. & B.)	*A. gossypii*	Cotton, *G. hirsutum*
17.	*Coelophora* sp.	*A. craccivora*	Cowpea, *Pisum sativam* L.
18.	*Coelophora bissellata* (Mul.)	*A. gossypii*	Cotton, *G. hirsutum*
19.	*C. orientalis* (S. & B.)	Aphid, *D. compositae*	Composite garden plant, *Gallardia pulchella* (F.)
20.	*Curinus coeruleus* (Mul.)	Psyllids	Subabul, *Luceana retusa* (Brew.)
21.	*Cryptolaemus monotiouzieri* (Mul.)	*Maconellicoccus hirsutus*, (Green), *Pseudococcus* spp.	Mulberry, *M. alba*, Sapota *Achras zapota*,
		Chloropulvinana spp.	Guava, *P. guajava*, Mango, *M. indica*
22.	*Harniolus mirajensis* (S. & B.)	Aphid, *D. compositae*	Sunflower, *H. annus*
23.	*Harmonia* sp.	Sorghum aphid, *R. maidis*	Sorghum *S. vulgare*
24.	*H. soyabini* (S. & B.)	Ground aphid, *A. craccivora*	Soybean, *Glycine soja* (Linn.)
25.	*H. octomaculata* (Fab.)	Aphids	Vegetable, garden plant
26.	*Hyperaspis maindroni*	Cotton aphid, *A. gossypii*	Cotton, *G. hirsutum*
27.	*Jauravia* (*Clanis*) *pallidula* Mots.	Scale insect	Bamboo, *B. bambu*, Teak *Tectonia grandis* L.

Contd....

Table 14–contd....

Sr. No.	*Predators*	*Pests*	*Host Plant*
28.	*J. soror*	Scale insect	Sapota, *A. zapota*, Guava, *P. guajava*
29.	*Illeis darbari* (S. & B.)	Sorghum aphid, *R. maidis*	Jowar, *S. vulgare,* Salad, *Spinacia oleracea* L.
30.	*I. sathei* (S. & B.)	Aphids	Sauf, *Foreniculuns vulgare* (Mill.)
31.	*Menochilus kharipi* (S. & B.) *M. shivajensis* (S. & B.)	Sorghum aphid, *R. maidis* *R. maidis*	Maize, *Z. mays* Jowar, *S. vulgare,*
	M. sexmaculatus	*R. maidis* *Aphis sacchari* Z.	Maize, *Z. mays* Sugar cane, *Saccharum officinarum* L.
		A. rumucis *A. craccivora* *Periqrinus maidis* A.	Jowar, *S. vulgare*
32.	*M. tuljapurensis* S. & B.	Aphid, *D. compositae*	Sunflower *H. annus* L.
33.	*Nephus* sp.	*Saccharicoccus saccari* (Cock.)	Bamboo, *B. bambu*
34.	*Pullus xerampelinus* (Mul.)	Cotton aphid, *A. gossypii*	Cotton, *G. hirsutum*
35.	*P. elegans*	*Pulvinaria* spp. *Pseudococcus* sp.	Mulberry *M. alba*
36.	*Pharoscymnus horni* (Weise)	*S. sacchari*	Sugar cane, *S. officinarum*
		Diaspri echninocacti	Mulberry, *M. alba* Red gram, *C. cajana*
37.	*Rodolia ruficollis* (Mul.)	Mealy bug	-
38.	*Rodolia* sp.	*Pseudococcus* sp.	-
39.	*Scymnus guimeti* (Mul.)	*A. malvoides*, Cotton aphid, *A. gossypii*	Malvacae plant

Contd....

Table 14–contd....

Sr. No.	*Predators*	*Pests*	*Host Plant*
40.	*S. nubilis* (Mul.)	Aphid, *Rhopalosiphum* sp.	Bajra *Pennisetum typhoides*
41.	*S. (pullus) marathi* S. & B.	Aphid, *D. compositae*	Sunflower, *H. annus* Jowar, *S. vulgare*
42.	*S. indica* (S. & B.)	*D. compositae*	Sunflower, *H. annus*
43.	*S. coccivora* (Diyar)	*Aphis* sp.	Teak, *T. grandis* Mulberry, *M. alba*
44.	*Sticholotis criberellata* (Fair)	Hard scales	Bamboo, *B. bambu* Sapota, *A. zapota*
45.	*Vernia polyphagi* (S. & B.)	Aphid, *R. maidis*	Jowar, Sorghum-*vulgare*
(b) Carabid Beetles			
46.	*Anthia sexguttata* Horn.	Grasshopper, *Heiroglypus banian* (Fab.)	Rice, *Oryza sativa*, Wheat, *Triticum* spp., Jowar, *S. vulgare*, Grasses *Grammia* sp.
47.	*Cicindela campestris*	Rice bug *Leptocorisa varicornis* (Thunberg)	Rice, *O. sativa*
48.	*Cicindela sexpunctata*	Rice bug *L. varicornis*	Rice, *O. sativa*
49.	*Chlaenius rayotus* (Bates)	Gram pod borer, *Helicoverpa armigera* (Hubn.)	Chick pea, *Cicer arietinum*, Tomato, *Lycopersicon esculentum*, Cotton, *G. hirsutum*, Jowar, *S. vulgare*
50.	*Dicranoncus* sp.	Rice bug, *L. varicornis*	Rice, *O. sativa*

Contd....

Table 14–contd....

Sr. No.	*Predators*	*Pests*	*Host Plant*
51.	*Parena laticineta* (Bates)	Black Headed caterpillar, *Nephantis scrinopa* Meyrick	Coconut *Cocos nucifera*
52.	*P. nigrolinoata* (Chand)	Tobacco caterpillar *Sodoptera litura* Fab.	Tobacco, *Nicotiana tabacum*
53.	*Pheropsophus hilaris* (Dej)	Rhinoceros beetle, *Oryctes rhinoceros* L.	Coconut, *C. nucifera*
54.	*Phlaeoderomius nigrolineatus* (Chand.)	Black headed caterpillar, *N. serinopa*	Coconut, *C. nucifera*
55.	*Tetragonoderus* sp.	Grasshopper, *H. banian*, Small Insects	Rice, *O. sativa*, Grasses *Graminia* sp.
(B) Order - MANTOIDEA			
(a) Mantids			
56.	*Choeradolis rhombicollis*	Caterpillars	
57.	*Creobroter urbana*	Leaf hoppers, *Nephotettix* spp.	Rice, *O. sativa*
58.	*Eremoplana* sp.	Grasshopper, H. banian	Rice, O. sativa
59.	*Hierodula bipapilla* S.A.	Mulberry silk worm, *Bombyx mori* L. Tasar silk worm, *Antheraea mylitta* D.	Mulberry *M. alba*
60.	*Humberticella* sp.	Grasshoppers, small insects	-
61.	*Mantis religiosa* L.	Rice blue Leaf hoppers *Typhlocyba maculifrons* Motsch.	Rice *O. sativa*

Contd....

Table 14–contd....

Sr. No.	*Predators*	*Pests*	*Host Plant*
62.	*Paratenodera sinensis*	Flies, small insects, grasshoppers	-
63.	*Schizocephala* sp.	Butterflies, small insects	-
64.	*Sphodromantis* sp.	Leaf hoppers, caterpillars	-
65.	*Stagmomantis carolina*	Butterflies, small insects	-
66.	*Tenodera* sp.	Rice leaf hoppers, grasshoppers	-

References

Ayyar, T.V.R. 1924. An undescribed Coccinellid beetles of economic importance. *J. Bombay Nat. Hist. Soc.,* **30**: 491-493. (W.L. 25676).

Andrews, H.E., 1929. The fauna of India including Ceylon & Burma, Coleoptera, Carabidae, Vol. **I**. Carabinae, pp. xviii + 431, Taylor & Francis, London.

Andrews, H.E., 1930. Catalogue of Indian Insects. Part **18** - Carabidae, pp. xxii + 388, Central Publication Branch, Calcutta, Govt. of India.

Andrews, H.E., 1935. The fauna of British India including Ceylon and Burma, Coleoptera, Carabidae Vol. **II** -Harpainae -1, pp. xvi + 323, Taylor & Francis, London.

Anonymous, 1972. The biological control of Cucumber pests. Glasshouse Crops. Res. Inst. Growers Bull. **1**. pp. 1-52.

Arnett, R.H., 1963. The beetles of the United States. p. 1112, Washington.

Ball, G.E., 1979. Consepts of carabid classification: History, holomorphology and higher taxa. In Carabid Beetles: Their Evolution, Natural History and Classification ed. T.L. Erwin, G.E. Ball, D.R. Whitehead, **1**: 63-111. Boston/London/The Hague: Jank. p. 635.

Beier, M. 1930. New and rare mantoidea (Orthoptera) in the British Museum *Ann. Mag. Nat. Hist.,* **(10)6**: 432-460.

Beier, M. 1941. New and scltene Mantodeen aus deutschen Museen. *Ann. Naturl. Mus. Wien.* **52**: 126-154.

Beier, M. 1963. New and bemerkenswerte mantiden Verschiedener Herkunft. *Stuttg. Beitr. Naturkd.* **106**: 1-11.

Beier, M. 1964. Blattopteroidea. Ordnung mantodea. *Bronn's Kl. Ordn. Tierreich* **(5) (3)6**: 849-970.

Bess, H.A.; Spurr, S.H.; Littlefield, E.W. 1947. Forest site conditions and the gypsy moth. Harvard Forest Bull. p. 22.

Bhaskar, H. 1992. Systematic and Ecological studies on predaceous Coccinellid beetles (Coleoptera: Coccinellidae). M.Sc. (Thesis) Dept. of Agricultural Entomology, U.A.S. Bangalore pp. 1-160.

Bhosale, Y.A., 1994. Studies on Indian Coccinellidae with special reference to predaceous species. Ph.D. Thesis, Shivaji University, Kolhapur. pp. 1-264.

Bolivar, I. 1897. Les Orthopteres de st. Joseph's College. a 'Trichinopoly. *Ann. Sac. ent. France,* **66**: 303-316.

Booth, R.G. 1991. A new genus and species of Coccinellid (Coleoptera: Coccinellidae) predator on *Rostrococcus* spp. (Hemiptera: Pseudococcidae) from India. *Bull. Ent. Res.,* **81**: 147-150.

Brown, W.J., 1962. A revision of the forms of *Coccinella* L. occurring in America north of Mexico (Coleoptera: Coccinellidae). *Can. Ent.,* **94**: 785-808.

Buckner, C.H. 1966. The role of vertebrate predators in the biological control of forest insects. *Ann. Rev. Entomol.,* **11**: 449-470.

Burmeister, H. 1838. Handbuch der Entomologie, Vol. **2** (Berlin). pp. 1-120.

Canepari, C. 1986. On some Coccinellids of Northern India and Nepal in the Geneva Museum of Natural History (Coleoptera: Coccinellidae). *Revue Suisse Zool.,* **93**: 21-36.

Carlson, R.E., Chiang, H.C., 1973. Reduction of an *Ostrinia nubilalis* population by predatory insects attracted by sucrose sprays. *Entomophaga,* **18**: 205-211.

Casey, T.L. 1899. A revision of the American Coccinellidae. *J. N. Y. Ent. Soc.,* **7**: 71-169.

Casey, T.L. 1908. Notes on the Coccinellidae. *Can. Ent.*, **40**: 393- 421.

Chapin, E.A. 1965. The genera of the *Chilocorini* (Coleoptera: Coccinellidae) *Bull. Mus. Comp. Zool.*, **133**: 227-271.

Chapin, E.A.; 1965. Coleoptera: Coccinellidae. *Ins. Micronesia,* **16**: 189-254.

Chapin, J.B., 1974. The Coccinellidae of Louisiana (Insecta: Coleoptera) *Agri. Exper. Sta. Doyle Chambers, Bull.,* **682**: 1-87.

Chapius, M.F. 1876. In Lacordaire, Nature des Insects et Genera des Coleopteres, Vol.12, Families des Erotyliens, des Endomichides, et des Coccinellides Paris. (In French).

Coppel, H.C. and Mertins, J.W., 1977. Biological Insect Pest Suppression. *Adv. Series Agri. Sci.,* **4**: 1-301.

Costa, A, 1849. Fanua del Regno di Napoli. *Coleoterri, Napoli, Sautto*. p. 112

Crotch, G.R., 1873. Revision of the Coccinellidae of the United States. *Trans. Amer. Entomol. Soc.,* **4**: 363-382.

Crotch, G.R., 1874. A Revision of the Coleopterous family Coccineillidae. pp. 311, London (I, II).

Darlington, P.J., 1943. Carabidae of mountains and islands: data on the evolution of isolated faunas and on attrophy of wings. *Ecol. Mongr.,* **13**: 37-61.

DeBach, Paul, 1964. Biological Control of insect pests and weeds. pp. 1-243. Chapman and Hall Ltd., London.

Diraviam, J. and Virakatmath, C.A., 1991. Population Dynamics of the introduced lady bird beetle, *Curinus coeruleus* Mulsant in relation to its psyllid prey *Heteropsylla cubana* Crawford in Bangalore. *J Biol. Control,* **4(2)**: 99-104.

Diraviam, J. and Virakatmath, C.A., 1991. Biology of the introduced lady bird beetle, *Curinus coeruleus* Mulsant (Coleoptera: Coccinellidae). *J. Biol. Control,* **5(1)**: 14-17.

Dobzhansky T., 1924. Die Palaarktischen Arten der Gattlung Coccinellia L. *Rev. Russ. Eng.,* **20**: 16-32.

Dobzhansky T., (1924). Dir Weiblichem Gener at jonsorgane der, Coccitn eliden als Artmerkmal hetrachtet (Col.) *Ent. Mitt.,* **13**: 18-27 (In German).

Dutt, G. R., 1927. Aphids and lady bird beetles. *Agric. India*, **22(4)**, 291-292.

Dyandechka, N.P., 1954. Coccinellids of the Ukrainian SSR. p. 156 Keiv (in Russia).

Economic Survey of Maharashtra, 1991-92. pp 13-20. Directorate Statistical Department Maharashtra State.

Emden, H.F. Van, 1965. The role of uncultivated land in the biology of crop pests and beneficial insects. *Sci. Hort.*, **17**: 121-136.

Fabricius. J.C., 1975. Systems Entomologiae, Lipsiae, p. 832.

Franz, J.M., 1961. Biologische Schadlingsbekampfung. In: Handbuch der Pflanzenkrankheiten. Richter, H. (Ed.) Berlin- Hamburg: Paul Parey, Vol. **VI**, pp. 1-302.

Ganglbaver, L., 1899. Die Kafer Von Mitteleuropa Band 3, pp. 1044. Wein (In German).

Ghani, M.A., 1962. A note on the identity of some species of genus *Ballia* (Coleoptera: Coccinellidae) *Proc. R. Ent. Soc. London (B)*, **31**:7 -8.

Ghorpade, K.D., 1974. Description of a new species of *Cryptogonus* Mulsant from Bangalore Southern India. (Coleoptera: Coccinellidae). *Oriental Ins.*, **8**: 55-60.

Ghorpade, K.D. 1977. A new species of *Pseudoscymnus* (Coleoptera: Coccinellidae) predaceous on Coconut scale, in Peninsular India. *J. Bombay Nat. Hist.*, **11**: 465-469.

Giglio-Tos, E. 1912. (Volume date 1911). Mantidi Esotici. *Ball. Soc. Entomol. Ital.*, **43**: 1-167.

Giglio-Tos, E. 1915. (volume data 1914). Mantidi Esotici. *Ball. Soc. Entomol. Ital.* **46**: 31-108.

Giglio-Tos, E. 1927. Mantidae *Das Tierreich* 50 Lief. (Berlin).

Glen, R. 1954. Factors that affect insect abundance. *J. Econ. Entomol.*, **47**: 398-405.

Gordon, R.D. and Chapin, E.A., 1983. A revision of the New World species of *Stethorus* Weise (Coleoptera: Coccinellidae). *Trans. Am. Ent. Soc.*, **109**: 229-276.

Gordon, R.D., 1978. West Indian Coccinellidae II (Coleoptera): Some scale predators with keys to genera and species. *Coleopterists Bull.*, **32**: 205-218.

Gordon, R.D., Pakluk, J., Slipinski, S.A. 1989. Carinodulini, a new tribe of Sticholotidinae based upon a new genus and species from Mexico (Coleoptera: Coccinellidae). *Coleopterists Bull.*, **43**: 359-364.

Graham, S.A. 1928. The influence of small mammals and other factors upon the larch sawfly survival. *J. Econ. Entomol.*, **21**: 301-311.

Greenslade, P.J.M., 1964. The distribution, dispersal and size of a population of *Nebria brevicollis* (F.) with comparative studies on three other Carabidae. *J. Anim. Ecol.*, **33**: 311-33.

Greenslade, P.J.M., 1968. Habitat and altitude distribution of Carabidae (Coleoptera) in Argyll, Scotland. *Trans. R. Ent. Soc. Lond.*, **120(2)**: 39-54.

Hagen, K.S., Franz J.M., 1973. A history of biological control In: History of Entomology. Smith, R.F., Mittler, T.E., Smith, C.N. (eds.) Palo Alto, Calif.: Annual Reviews, pp. 433-476.

Hambleton, E.J., 1944. *Heliothis virescens* as a pest of cotton, with notes on host plants in Peru. *J. Econ. Entomol.*, **37**: 660-666.

Hartman, R.C. 1971. Biological fly control spreads. Poultry Dig. **30**: 224-227.

Hodek, I. 1973. Biology of Coccinellidae. Academic Publishing House, Czech pp. 1-77.

Hodek, I., 1973. Biology of Coccinellidae. Academia, Prague, pp. 1-260.

Holdren, J.P., Ehrlich, P.R. 1974. Human Population and the global environment. *Am. Scientist*, **62**: 282-292.

Horn, W., 1915. Genera Insectorum, Fasc. 82A-C, Coleoptera. Adephaga. family Carabidae, subfamily Cicindelidae. Brussels: Desmet-Verteneuil. 486 p.

Ipetri, G.A. 1965. Contribution a letude de la specificite chez les Principales Coccinelles aphidiphages des Alpes-Maritimes et. des. Basses. Alpes, *Ibid*, **10**: 159-78.

Kamiya, H., 1961. A revision of the tribe Scymnini from Japan and Loochoos (Coleoptera: Coccinellidae). *J. Fac. Agr. Krushu Univ.*, **11**: 275-302.

Kamiya, H., 1965. Comparative morphology of larvae of the Japanese Coccinellidae, with special reference to the tribal phylogeny of the family (Coleoptera). *Mem. Fac. Lib. Fukui Univ.,* **14(2)**: 83-100.

Kapur, A.P., 1947. A revision of the tribe Aspidimerini Weise (Coleoptera: Coccinellidae). *Bull. ent. Res.,* **39**: 297-320.

Kapur, A.P., 1948. On the old world species of the genus *Stethorus Weise* (Coleoptera: Coccinellidae). *Bull. Ent. Res.,* **39**: 297-320.

Kapur, A.P., 1948. On the Indian species of *Rodolia* Mulsant (Coleoptera: Coccinellidae). *Bull. Ent. Res.,* **39**: 531-538.

Kapur, A.P., 1950. The biology and external morphology of the larvae of Epilachinae. *Bull. Ent. Res.,* **41**: 161-208.

Kapur, A.P., 1954. Systematic and biological note on the San Jose scale in Kashmir with description of a new species (Coleoptera: Coccinellidae). *Rec. Indian Mus.,* **52**: 257-274.

Kapur, A.P. and Rao, S. V., 1962. A new species of ladybird beetle (*Adalia sirmnondsi* sp. n.) (Col: Coccinellidae) feeding on *Adelges* (*Chermes*) spp. (Hem.: Adelgedae) on silver fir (*Abies* sp.). *Entomophaga*, **7**: 47-51.

Kapur, A.P., 1963. The Coccinellidae of the third Mount Everest Expedition, 1924 (Coleoptera). *Bull. Brit. Mus. (Nat. Hist.), Ent.,* **14**: 1-48.

Kapur, A.P., 1967. The Coccinellidae (Coleoptera) of the Andamans. *Proc. Nat. Inst. Sci. India,* **32(B)**: 149-189.

Kapur, A.P., 1969. On some Coccinellidae of the tribe Telsimiini with description of new species from India. *Bull. Syst. Zool., Calcutta,* **1(2)**: 45-56.

Kapur, A.P. and Munshi, S., 1970. A new species of *Scymnus* (*Pullus*) Mulsant, *S.* (*P*) *nymphaeus* sp. nov. (Coleoptera: Coccinellidae), Predaceous on the aphid *Rhopalosiphum nymphaeae* (Linn.) in Calcutta, India. *Rec. Zool. Survey India,* **62**: 1-4.

Kapur, A.P., 1972. The Coccinellidae (Coleoptera) of Goa. *Rec. Zool Survey India,* **66**: 309-320.

Kirby, W.F. 1904. 'A' synonymic: Catalogue of Orthoptera', Vol. **I** (London).

Kirby, W.F., 1904. A synonymic catalogue of Orthoptera *Cat. Orth. Brit. Mus.* **1**: 210.

Korschefsky, R., 1931. Coccinellidae I. Coleopterorum Catalogues, Pars, **118**, Berlin, p. 224.

Korschefsky, R., 1932. Coccinellidae II. In Junk-Schenkling *Coleopterorum Catalogues, Pars,* **120**, Berlin, Junk, p. 433.

Korschefsky, R., 1932. Coleopterorum Catalogues Pars, **118**. Coccinellidae I, Berlin, pp. 224.

Korschefsky, R., 1933. Entomological investigations on the spike Disease of Sandal. (16) Coccinellidae (Coleoptera). *Indian Forest records,* **19(6)**: 1-17.

Kugelann, J.G., 1794. Verzeichniss der in einigen Gegenden Preussens bis jetzt entdeckten Kaferarten nebst kurzen Nachrichten von denselben. *Neues Magazin Liebb. Ent. Heft.,* **5**: 513-582.

Lawson, F.R., Rabb, R.L., Guthrie, F.E., Bowery, T.G.; 1961. Studies of an integrated control system for hornworms on tobacco. *J. Econ. Entomol.,* **54**: 93-97.

Leach, W.E., 1815. (Articles on entomology). In Brewster, Edinburgh. *Encyclopaedia* **9**: 57-172.

Lefebvre, M.A. 1835. Nouveau groupe d orthopteres de la Famille des Mantides. *Ann. Soc. Ent. France,* **4**: 499.

Lefory, H.M. 1909. Indian Insect Life. TTPP, New Delhi.

Lewis, G., 1873. Notes of Japanese Coccinellidae. *Ent. Mon. Mag.,* **10(2)**: 54-56.

Lewis, G., 1879. On certain new species of Coleoptera from Japan. *Ann. Mag. nat. Hist. Londi.,* **5(4)**: 459-467.

Linnaeus, C., 1758. Systema naturae per regna tria naturae, Secundum Classes. Ordines, genera, species, cum Characteribus, differentiis, Synonymis locis. Ed. **10(1)**, Holmiae, pp. 1-823.

Londorth, C.H., 1969. The ground-beetles of Canada and Alaska. Part-I *Opus. Ent. Suppl.* **35**: I-XLVIII.

Mader, L., 1926-37. Evidenz der palaarktischen Coccinelliden und ihrer Abberrationen. Wort und Bild I. Teil., 412+12.

Mader, L., 1941. Coccinellidae, I Teil. *Explor. Parc. nat. Albert. Miss. de Writte fasc.* **34**, p. 136.

Mader, L., 1953. Evidenz der palaarktischen Coccinelliden und ihrer Abberrationen. Wort und Bild I. *Teil. Engom. Arb. Mus. frey.*, **6**: 764-1035.

Mader, L., 1954. Coccinellidae, III Teil. *Explor. Parc. nat. Albert. Miss. de Writte fasc.* **80**: 206.

Manjunath, T.M. 1992. Biological Control from Lab to land: Roll of Commercial insectaries. *Proc. Nat. Workslwp Biotech Appro. Bio. Control Phytophaques insects*, pp. 32-33.

Mayne, W.W., 1953. *Cryptolaemus montrouzieri* Mulsant in South India. *Nature*, **172**: 185.

Miyatake, M., 1967. Notes on some Coccinellidae from Nepal and Darjeeling District of India (Coleoptera). *Trans. Shikoku Ent. Soc.*, **9(3)**: 69-78.

Miyatake, M., 1969. The genus *Plotina* and related genera (Coleoptera: Coccinellidae) *Pacific Insects*, **11(12)**: 197-216.

Miyatake, M., 1970. The East-Asian Coccinellid-beetle preserved in the California Academy of Sciences, tribe Chilocorini. *Mem. Coil. Agr. Ehime Univ.*, **14(2)**: 303-340.

Miyatake, M., 1985. Coccinellidae collected by Hokkaido University expedition to Nepal, Himalaya, 1968 (Coleoptera) *Insecta Matsumarana*, **30**: 1-33.

Moore, B.P. 1963. Studies on Australian Carabidae (Coleoptera) 3 - The *Psydrinae. Trans. R. Ent. Soc. Lond.* **115**, PT 11.

Mukherjee, T.K. and A.K. Hazra, 1993. Insecta: Mantodea, State Fauna Series 3: Fauna of West Bengal, Part **4**: 475-510.

Mukherjee, T.K. and A.K. Hazra, 1985. Insecta: Mantodea, *Rec. Zool. Surv. India,* **82 (1-4)**: 33-39.

Mukherjee, T.K. and A.K. Hazra, 1985. New record of a deserticolous Mantid family (Mantodea: Eremiaphilidae) from India, *Entomon.* **10(3)**: 245-247.

Mukherjee, T.K. and A.K. Hazra, 1985. On some little known and a new praying mantids (Mantodea) from Mulla-Periyar Tiger Reserve, Kerala, India, *Entomon.* **10(4)**: 255-259.

Mukherjee, T.K. and Hazra, A.K. 1983. On a small collection of Mantidae (Dictyoptera) from Maharashtra, India, with the description of a new species, *Rec. Zool. Surv. India*, **80**: 459-465.

Mulsant, M.E., 1850. Species des Coleopteres trimeres Securipalpus, p. 1104.

Mulsant, M.E., 1866. Monographic des Coccinellides Ire partie Coccinelliens, pp. 1-294, Paris.

Nagraja, H. and Hussainy, S.U., 1968. A study of six species of *Chilocorus* (Coleoptera: Coccinellidae) predaceous of San Jose and other scale insects. *Orient Insects*, **1**: 239-241.

Nakahara, L., Magamine, W., Matayoshi, S. and Kumashiro, B., 1987. Biological control programme on the leucaena psyllid, *Heteropsylla cubana* Crawford. (Homoptera: Psyllidae) in Hawaii. *Leucaena Res. Rpt.*, **7**: 39-44.

Nakane, T., 1959. Entomological results from the scientific survey of the Tokara Islands. (Coleoptera: Coccinellidae). *Sci. Rep. Kyoto. Pref Univ.*, **3**: 45-52.

Olivier, G.A. 1792. 'Encyclope, die methodique Vol. **7** (Paris).

Omkar and Bind, R.B. 1993. Records of Aphid natural enemies Complex of Uttar Pradesh. II. The Coccinelids. *J. Adv. Zool.*, **14(2)**: 96-99.

Pajni, H.R. and Singh, J., 1982. A report on the family Coccinellidae of Chandigarh and its supporting area (Coleoptera). *Res. Bull. Punjab Univ. Sci.*, **33**: 79-86.

Pajni, H.R. and Verma, S., 1985. Studies on the structure of the male genitalia in some Indian Coccinellidae (Coleoptera). *Res. Bull. Punjab Univ. Sci.*, **36**: 195-201.

Paroda, R.B., 1993. Crop science research for sustainable agriculture in India. *Proc. Nat. Acad. Sci. India*, **63(B)**: 97-114.

Patro, B. and Behera, M. K., 1992. Development and feeding potential of *Scymus* sp. on the Bean Aphid, *Aphis craccivora. J. Biol. Control,* **6(2)**: 64-66.

Pearson, David L., 1988. Biology of Tiger beetles, *Ann. Rev. Entoml.*, **33**: 123-47.

Pearson, R., 1964. Alary polymorphism in *Benobidion grapei* Gyll (Col., Carabidae). Entomologists. *Mon. Mag.*, **100**: 152-3.

Plaut, H.N. 1965. On the phenology and control value of *Stethoru punctillum* Weise as a predator of *Tetranychus cinnabarinus* Boisd. in Israel. *Entomophaga*, **10**: 133-137.

Pollard, E. 1968. The effect of removal of the bottom flora of a hawthorne hedgreow on the Carabidae of the hedge bottom. *J. Appl. Ecol.*, **5**: 125-139.

President's Science Advisory Committee, 1965: Restoring the Quality of our Environment Report of the Environmental Pollution Panel. Washington, D.C. The White House Government Printing Off.

Putman, W.L., 1957. Laboratory studies on the food of some coccinellids (Coleoptera) found in Ontario peach orchards. *Can. Ent.*, **89**: 572-579.

Puttarudraiah, M., and Channabasavanna, G.P., 1952. Discovery of *Cryptolaemus nwntrouzieri*, Mulsant (Coccinellidae: Coleoptera) in Bangalore, South India, *Nature*, **169**: 377-378.

Puttarudraiah, M., and Channabasavanna, G.P., 1953. Beneficial Coccinellids of Mysore-I. *Indian J. Ent.*, **15**: 87-96.

Puttarudraiah, M., and Channabasavanna, G.P., 1955. Some beneficial Coccinellids of Mysore-II. *Indian J. Ent.*, **54**: 156- 159.

Puttarudraiah, M., and Channabasavanna, G.P., 1956. Beneficial Coccinellids of Mysore-III. *Indian J. Ent.*, **17**: 1-5.

Rabb, R.L., Stinner, R.E. Carlson, G.A. 1974. Ecological principles as a basis for Pest management in the agroecosystem. In Proceedings of Summer Institute on Biological Control of Plant Insects and Diseases Maxwell, F.G., Harris, F.A. (eds.) *Jackson Univ., Mississippi*, pp. 19-45.

Rafiq Ahmed, 1973. A new tribe of the family Coccinellidae (Coleoptera). *Bull. ent. Res.*, **62**: 449-452.

Rao, C.S., Rao. S.V. and Rao, B.H., 1981. Biology of the exotic preadator *Chilocorus cacti*, (Coccinellidae: Coleoptera) on sugarcane scale Insect *Melanaspis glomerata* Green. *Cooperative sugar*, **12(6)**: 303-304.

Rao, T.V. and David, A.L., 1958. The biological control of Coccid. pest in South India by the use of the beetle, *Cryptolaemus montrouzieri* Muls. *Indian J. Agri. Sci.,* **28(4)**: 545-552.

Rao, V.P., 1971. A review of the biological control of insects and other pests in South East Asia and the Pacific region. Commonwealth Agricultural Bureau, Farnham Royal, Slough SL2 3BN, England, pp. 1-149.

Rawat, U.S., Sangal, S.K. and Pawar, A.D., 1992. Biology of *Chilocorus bijugus* Mulsant (Coleoptera: Coccinellidae), Preadator of San Jose scale, *Quadraspidiotus perniciosus* (Comstock). *J. Biol. Control,* **6(2)**: 97-100.

Redtenbacher, L., 1843. Tentamen dispositionis generum et specierum Coleopterorum Pseudotrimerorum Archiducatus Astriae, reprint 1844. *Zeitschr. Entomol.,* **5**: 113-132.

Saha, S.K., Mukherjee A.K. & Sengupta T. 1992. Carabidae (Coleoptera: Insecta) of Calcutta. Records of the Zoological Survey of India, pp. 1-63.

Sasaji, H., 1967. On the Coccinellidae attacking the aphids in Japan and Ryukyus. *Mushi,* **40**: 1471-75.

Sasaji, H., 1968a. Phylogeny of the family Coccinellidae (Coleoptera). *Etizenia,* **35**: 1-37.

Sasaji, H., 1968b. Description of the Coccinellid larvae of Japan and the Ryukyus (Coleoptera). *Mem. Fac. Edn. Fukui Univ.,* **2(18)**: 93-136.

Sasaji, H., 1970. Biogeographical consideration of the Coccinellidae of the Tsushimas, Japan (Coleoptera). *Bull. Nat. Sci. Mus. Tokoyo.* **13**: 206-217.

Sasaji, H., 1971. Fauna Japonica Coccinellidae (Coleoptera), Academic Press, Japan, p. 340.

Sathe, T.V., 1992. Fauna of aphids on plants of economic importance found in Western Maharashtra, *India. J. Curr. Biasci.,* **9(1)**: 27 -31.

Sathe, T.V., 1986. Biology of *Catesia diumii* R and N. (Hym.: Braconidae) a larval parasitoid of *Exelastis atomosa* Walsingham. *Oikaassay,* **3**: 31-33.

Sathe, T.V., 1998. Sericultural Crop Protection. Asavari Pub., pp. 1-120.

Sathe, T.V. & Bhoje, P.M., 2000. Biological Pest Control. Daya Publishing House, pp. 1-122.

Sathe, T.V. and Bhosale Y.A., 2001. Insect Pest Predators, Daya Publishing House, Delhi, pp. 1-169.

Sathe, T.V. & Jadhav, A.D., 2001. Sericulture and Pest Management, Daya Publishing House, pp. 1-195.

Sathe, T.V. & Margaj, G.S., 2001. Cotton Pests and Biocontrol Agents. Daya Publishing House, pp. 1-161.

Saussure, H. de 1871. Melanges Orthopterologiques. Fasc. 3 *Mem. Sac. Phys. Hist. Nat. Geneve,* **21**: 1-214.

Saussure, H. de, 1869. Essai d'un systema des mantides, *Mitt. Schweiz. Entamol. Ges.,* **3**: 49-73.

Saussure, H. de. 1870. Additions au systeme des Mantides. *Mitt. Schweiz, Entomol. Ges.,* **3**: 221-244.

Serville, J.G. Audinet (1838: title page date 1839). Histoire Naturelle des Insects' Orthopteres. (Paris).

Sicard, A., 1907. Coleopteres Coccinellides du Japon. *Bull. Mus. Hist. Nat. Paris.,* pp. 211.

Sicard, A., 1909. Revision des Coccinellides de la faune. *Ann. Soc. ent. Fr.,* **78**: 68-165.

Simmonds, F.J. 1970. Biological Control. Part 4/8. Sampod Press Ltd., Bangkok.

Singh, A. and Sadhu, A.N., 1986. Agricultural problems in India. Jammu University, Jammu, p. 386.

Smith, R.F. 1969. Integrated control of insects a challenge for scientists. *Agri. Sci. Rev.,* **7(1)**: 1-5.

Southwood T.R.E., Way, M.J., 1970. Ecological background to Pest Management In: Concepts of Pest Management. Rubb, R.L., Guthrie, F.E. eds.), Rateigh: North Carolina st. univ., pp. 6-29.

Stal, C. 1813. 'Representation Exactment des Blattes' (Amsterdam).

Stal, C. 1871. Orthoptera quoedam africana, *Ojvers. K. Vetensk Akad. Fork.* **3**: 375-401.

Stal, C. 1877. Systema Mantodeorum Bin. *K. Svenska Vetnsk Akad. Handl.,* **4(10)**: 1-91.

Subramaniam, T.V., 1923. Some coccinellids of South India. *Rep. Proc. Fifth Ent. Meeting,* Pusa, pp. 108-118.

Subramaniam, V.K., 1954. Control of the fluted scale in Peninsular India. *Indian J. Ent.,* **15(1)**: 103-120.

Subramaniam, V.K., 1955. Control of the fluted scale in Peninsular India. *Indian J. Ent.,* **16**: 39-415.

Sweetman, H.L., 1958. The principles of Biological control. Dubuque; *Wm C. Brown.,* p. 8.

Thunberg, C.P. 1784. Novae Insectorum Species. Dissertationes entomologicae Vol. **3** (Upsaliae).

Timberlake, P.H., 1943. The Coccinellidae or lady beetles of the Koebele Collection Part II, Hawaii. *Plant Rec.,* **47**: 1-67.

Usman, S. and Puttarudriah, M., 1955. A list of the insects of Mysore including the mites. *Ent. Ser. Bull.,* **16**, Dept. Agric,. pp. 1-194.

Van Vreden G. & Ahmadzabidi A.L., 1986. Pests of Rice and their natural enemies in peninsular Malaysia. Pudoc Wageningen. pp. 110-151.

Watson, W.Y., 1956. A study of the phylogeny of the genera of the tribe Coccinellini (Coleoptera). *Centr. Roy. Ont. Mus. Toronto (Zool.),* **42**: 1-52.

Way, M.J., 1954. Studies on the association of the ant *Oecophylla longinoda* (Latr.) (Formicidae) with the scale insect *Saissetia zanzibarensis* Williams (Coccidae). *Bull. Pest Control. Res.,* **45**: 113-134.

Werner, F. 1931. Further notes on Indian Mantids or praying insects. *Proc. Zool. Soc.* London, pp. 1329-1334.

West Wood, J. 1889. Revisio Insectorum: Family Mantiderum: 52 + iii, 14 Pls. London.

Williams, 1904. *Trans. Ent. Soc., Lond.,* p. 125.

Wood-Mason, J. 1877. On a small collection of Orthopterous insects, four new species. *Ann. Mag. Nat. Hist.,* **20**: 74-77.

Wood-Mason, J. 1878. On new and little known mantidae. *Proc. Zool. Soc. London,* **38**: 580-587.

Wood-Mason, J. 1879. Description of *Sigerpes occidontalis*, the type of new genus of Mantodae from West Africa. *Proc. Soc. Beng.*, 257-259.

Wood-Mason, J. 1882. On new and little known Mantodae. *J. Asiat. Soc. Bengal.*, **51**: 21-36.

Wood-Mason, J. 1884. *Phyllathelys* a remarkable genus of Mantodae from Oriental Region. *J. Asiat. Sac. Beng.*, **3**: 1-20.

Wood-Mason, J. 1889. A Catalogue of Mantodea with description, Indian Museum, Calcutta No. I (Calcutta).

Wood-Mason, J. 1891. A Catalogue of Mantodea, Indian Museum Calcutta No.**2** (Calcutta).

Zaslavskij, V.J., 1965. Fam. Coccinellidae. (key to the Insects of European region of U.S.S.R.) **2**: 319-326. (In Russian).

Index

D

E

S

T